Note: Best file 5-18-99

AN EYEWITNESS ACCOUNT

OF THE NOVEMBER 11, 1992

HURRICANE INIKI

ON KAUAI'I
HAWAII
USA

" THE PIERCING WIND "
BEFORE AND AFTER

By ROLAND L. CLARK II

THE IRISH CATHOLIC REVOLUTION PUBLISHING CO.
A Royalist/Radical Publishing Co.

AN EYEWITNESS ACCOUNT
OF THE NOVEMBER 11, 1992

HURRICANE INIKI
Note: Best file 5-18-99

A Lost and Found Manuscript of Roland L. Clark II'S

The **HURRICANE INIKI** manuscript was given to Dr. Patrick A. O'Dougherty by Roland Clark II to preserve in case of a possible loss or theft. Dr. O'Dougherty's publishing company apartment was ransacked on July 6, 2002. Hence this publication. by Irish Catholic Revolution Publishing.

Royalty Rights
Roland Clark has royalty rights to this publication. Both Roland Clark II and Patrick A. O'Dougherty grew up in St. Cloud, MN. O'Dougherty met Clark at the Dorothy Day Center in St. Paul, MN. He got him an office at the Riverside Plaza Resource Center in Minneapolis where he replicated this manuscript which had been stolen in the Los Angeles airport. Both writers attended St. John's University in Collegeville which is located about five miles from St. Cloud, MN.

This book founds a reconciliation with the earth theology out of the Benedictine Oblate group out of St. John's University and a St. Cloud School of ecology writing

Final judgement on this hurricane and manuscript has not yet been reached.

This book is an Aloha manuscript from islands containing various indigenous peoples dialects.

ISBN: 0-9741978-0-7

COPYRIGHT 2004: IRISH CATHOLIC REVOLUTION PUBLISHING COMPANY

LIBRARY OF CONGRESS CATALOG CARD NUMBER:

PRINTED IN THE UNITED STATES OF AMERICA

DEDICATION

JIM and PAMELA MORRISON and THE DOORS

"THERE IS THIS WORLD AND OTHERS, IN BETWEEN ARE THE DOORS.".
"INTO THIS WORLD WE'RE THROWN LIKE RIDERS ON THE STORM".

- JIM MORRISON

TO THE BEST IN THE WORLD, THE HAWAIIAN ALOHA PEOPLE.

Wanda Woods, One of Kauai's best artists. So many needed your "ALOHA SPIRIT. We were the RIDERS AFTER THE STORM. You had a helping, caring and tolerant spirit for many weary hurricane survivors.

JUDY J. IRVING (Independent Documentary Film Group of SFO, CA.)
Thanks for a lot you taught, reach out and communicate.
If I ever understood you well,
" Those were the days my friend, these still are."

To "Di" Patricia for LOVE
A lifelong process of Healing, all learned in "Mendoland"." I Gave what I got until I had no more, was lied to and cheated trying ...just to even the score, You know I love you and furthermore, it was time to go, you had the open door". Thank your intemperance, the Lord and Hawaiian Air, you left.

To Irene, It must surely take a lot of patience to be a Mother, yet the best of life takes ...time.

To Gracie Slick "You are the crown of Creation and you've got no where to go, they
Cannot tolerate your mind!"
(lyric courtesy Jefferson Airplane}

Any resemblance to persons living or dead is coincidental or... a possible result of P.I.S.S. (POST-INIKI SHOCK SYNDROME) In this *novel* form some human inaccuracy is likely unavoidable for a number of reasons due to innumerable thieves of my manuscripts, typewriters, wordprocessors, discs and a... number of various other causes, not the least of which a lot of poverty and several years of harassment from ridiculous sources. Reader note this book was rewritten over three times due to theft and loss of original manuscript. Perhaps most noteworthy of many obstacles surmounted were the various difficult outdoor conditions under which this was written, many times, little protection from any elements that the great "PELE" could cast ashore at the author.

PREFACE and CREDITS

ONE great Hawaiian spirit, Bobby Kamaanuu of Honolulu. (in the late 1980's period) pointed the direction (misdirection's) of Oahu's urban development., "They are cutting down the fruit trees all over in Honolulu." His tone of voice was asking " Is this destruction of my homeland necessary". I see this as a very shortsighted, extremely shortsighted form of gain, you could draw a parallel with cutting of the Amazon forest.. to grow more cattle for MC Donalds. The few trees in Honolulu, even fewer in New York or any major city are deserving of nameplates, special identity number far more than trashing on a hillside dump The government agencies in action, are just too ignorant for any further consideration or mature an outlook, in HONOLULU for revenue and "promo" of COMMERCIAL developers. Is that any way to show respect for the very land they are developing ? The real Charm, Beauty of these islands is being destroyed at a rapid rate, as I most certainly witnessed on every island visited. More about all that in Chapter One. This made Bobby very depressed. As far as I could observe, he simply was never a really happy person again, such as the one I'd met only a few years earlier by "The Keyhole" a lunchtime rest- spot, on Dillingham Av. In NO. Honolulu. He was outgoing, happy, generous, kind, and very practical. I watched his health deteriorate, as I've seen many over- pressed Hawaiians. He worked too hard and later became disabled on the job. Three generations of his family made me feel welcome and I am sure his father (who played with Louis Armstrong) is now smiling down on Bobby from that big band up in the "TERRITORIAL AIRWAVES." Music has always been their best "meds."

IT'S INTERESTING to note within months after Bobby's statement about the situation developing in HI., with "lifestyle" getting in the way... on down to trees, his entire apt. complex was being "spied on" to keep extra visitors away. Later still, it was totally leveled, destroyed in order to build another one on the exact same Hawaiian lands at their expense. I liked their old apts. "Moh-betta." His move to Wailuku on Maui was mainly in (protection) consideration of the children. I can respect anyone for that set of values and witnessed the regular events of violence against my own person in Honolulu, and numerous friends including senior citizens. All this needless, connected with the drug traffic, various street and domestic violence, but worst of all; non-cooperation from the officials after theft of my (home) sailboat and all I then owned. Thus ended my visit of Oahu and the big HI. City.

TO THE many other beautiful, artistic and inspiring people in Hawaii, land of Aloha, I thank you for introducing me to a whole new way of life, being rapidly changed, inundated and constantly challenged by a tremendous "strictly from commercial" attitude to a way of life that is sacred to your people. Hundreds welcomed me and I shall never forget your many kindness.

To the People of Kauai'Is Island hospitality was felt , perhaps felt strongest on KAUAI and I'd say that's due to the strong family ties the Beauty of this place... it was incomparable to any other I'd visited during the more than hundred times I'd been around the world. The story of how I came to move there is one of your guidance, Robert K. and other friends who I never even got to know. But told me consistently a similar TALE.

" You would like it on Kauai' " the people and lifestyle, beauty, scenery, the No. West Shore etc. Bobby told me the same in his own way and also another Bob F., his family had real estate business on Kauai. I slowly developed curiosity and interest but in the end it was the violence and thievery of the inner city of Hon. which was getting monthly worse, to the ridiculous point when I couldn't replace my property as quickly as it was being stolen. Bobby and his family came to a similar conclusion as did the "Mainland" in the late "80s". Fences were going up, more cops, security devices, impediments, cameras, under-covers (some working as/ or impersonating "Couples") and finally more prisoners, prisons for the political prisoners and innumerable others. Finally the HI. Prisoners were being shipped out to the mainland, Texas; for which the total society is paying. It need not be mentioned in many ways and better left unsaid, yet we must *begin* to understand the roots of this. We all pay for this breakdown of personal trust when violence and dishonesty infiltrates any society core and mistrust replaces that binding thread of harmony. I watched the City and County of Honolulu become more tightly encircled with STEEL as month after month the beautiful Mango and Guava trees were cut, *replaced* by various bared lots; not so wisely done commercial developments and hurricane fences. In the fall of 1996 I recall bike-riding from Nimitz highway North for a six hour ride up the mountains, I couldn't find an open stretch (unfenced) place to even lay out a sleeping bag and rest. This Hawaii has certainly became the "Land of the Brave" if you are into that form. That trip ended early in the morning, painfully exhausted sleeping near some of the thousands of acres that the Navy, Army, Marine Corps, Air Force and Coast Guard reserves for our complex military.

THE MILITARY with tens of thousands of acres of "RESERVATIONS" and the tourist industry with its "visitor reservations" , are both HAWAII'S main "GROWTH" industry, then there's gambling. It's worth mentioning that on Kauai as other isles, the soil cannot grow sugarcane any longer because of primary soil depletion, another money factor to the growers is the incredible amount of water per acre required for sugar cane. This led to widespread coffee planting as of 1991 on Kauai and earlier on the Big Island.

PREFACE AND CREDITS

ANYWAY YOU look at it, Hawaii is certainly "Well Secured" on the Surface, but WE RESIDENTS know better. The HI. Guy on the No. shore of Kauai'I described these islands as "Armed to the teeth" but is it with the traditional and original spirit of Aloha, a flower wreath of blossoms and island hospitality for all visitors from all places? I'M NOT going to try and project any sad (Kapuu) images, Hawaii is a very happy place for tourists who visit maybe a week, also the landowners, those with the ways and means and a third smaller group, the creative artists, writers, musicians, for those who can attune and maintain harmony with nature. Between 1989 and just the brief three years until Iniki, there was so much short-sighted development change, bad looking change, the ruthless killing off of a wide spectrum of natural life forms by those without conscience.

Especially I am trying to project and share the unique images, eyewitness encounters seen over the total period of eight years visiting (especially the Pre and post Iniki) there was so much life and beauty in Kapaa and also lovely Hanalei Bay on the NO. shore of Kauai'i. Would I be fair to Bobby and Hawaii, in not voicing concern and conscience about this ongoing rape of Hawaii   FROM
FIRST morning awakening to this look at Hawaii as a new world, it was new adventure every day. At the far end of Sand Island rd. and beach, The First Hawaiian I spoke with, treated me with respect, encouragement, and generosity! His wife was smiling in her warm, gentle Hawaiian way, Island Style. They asked me to write an *ACTUAL ACCOUNT OF Real-Life in HAWAII* as contrasted with the sterilized & stylized "Ream-Team of Reporters" generating fantasy for Howles by which Mr. Michener made his mark in a very superficial capacity, for the fly-in fly-out audience. "Hawaii 5-0 was glorious to get people to see the picture(s) but again, another male, adult, chauvinistic, well moneyed, phony culture a/k/a/ English Aristocratic fantasy. Todays "quality" of life in Hawaii isn't a tourist package but a slowly degrading quality of Environment for all, the tourists, the resident Hawaiians and of course... the wildlife. Now would be as good a time as any to mention, Wildlife. More species of wildlife have been decimated (totally killed off, extinct) in HI., than all the other states put together.

As recently as "The Fifties" the majority of islanders working, frowned on excessive work and considered their family- LIFE a higher priority than money pursuit, both parents working was unheard of. Island-style life consisted of contact with the outdoors, a wellness based and very hospitable case of "Polynesian Paralysis," when island lore and Hawaiian sub-tropic music and love for the land -fever was far less commercial, more alive, On Kauai, this is still strong, perfumed with ginger blossoms, visions of the many island flowers blazing red hibiscus flowers for miles seen every day in this sub-tropic paradise. It's the essence, most sensitive and beautiful part of Hawaii forever lost to Oahu yet it remains more on Kauai' than any of the other isles I visited. "INIKI" brought out a lot in the real character of these people which during other " times" is simply submerged, like Hawaiian "Menehunes" working only at night.

Writing was a big one of half a dozen reasons to leave the "Mainland." Hawaii deserves THE credit, certainly an inspiring place for totally new input, completely opposite from living in the desert area, Tucson. There were some terrible city things and lessons to forget which I had to learn all over again in yet another city. The paradox of that statement is as Jim put it so eloquently "Learn to forget." The decision had been made to give away excess property, pack and leave, I simply knew no one to visit on any of the HI. Isles. Making new friends was in order. I was "a stranger in a very strange land". Let the significance of this understatement be learned by a new arrival with little means. To explore the HI. climate, its flora and fauna could have taken a LIFETIME or even... many, if a person had that kind of time "in Paradise". An expert camper could have spent years finding harmony in living out closer with nature.. I spent over six years at it and only scratched the surface with a finger nail fascination. The boy within was free all over again and this place, Hawaii especially Keehi Lagoon in Hon. was a new exploration trip everyday in an incredible land of diverse (people from eighty some national origins) and brilliant life forms above and below the Pacific waters, rainbows of color, depending on which bugs, fish, flowers gekkos and even the weather changes witnessed. I used to wonder why a rainbow is on the license plates, then I saw it shine (rainbow) and rain on and off, seven times in one day.

TO QUOTE Jack Griffith London in reference to Hawaii "These people don't know what they've got." That he observed a long time before the scheduled tourist flights began arriving in Keehi lagoon, Honolulu, HI. Island of Oahu which, coincidentally was my very first and last home on "The Gathering Place" as the Hawaiians refer to the island of Oahu. It's still a terrific biking place, one of the ultra modern cities on the planet where bikes are consistently quicker than cars, more practical, economical and a hell of a lot more fun..

A very bitter Hawaiian woman (KamaiianaWaihine) elder put it this way "They have taken our land and now are taking away Hi. homelands, about all we still have left ... is the climate!". At that time She'd been forced to live in *his* car with his semi-disabled wife. Several years later, sleeping in ones car in Hon. Was declared illegal and shortly after all that. The many Hawaiian People began to realize, it was an all-out war on survival of the poor .

PREFACE AND CREDITS
CONCLUDED

It's appropriate to mention and emphasize at this time, regardless of the TV imagery of Hollywood, that in fact most Natives of Hawaii, live in poverty well hidden (discarded) from the Media but readily visible, riding a bicycle through the industrial areas along Nimitz Hwy.. The real intent was/ is to drive out all who didn't generate revenue to the satisfaction of the land-controllers. That idea was related via a longtime island dweller.

IT SEEMS only a few yesterdays ago I glanced at a picture of "Diamondhead" postcard on a rough hewn oak desk in North Tucson, AZ , From that glimpse of a vision the basic MOTIVATION idea began to develop, Should I move to Hawaii? It was a daring move and I hesitated for about a year. WORK as a writer takes new inspiration, images and vision. On the "Mainland" there were tough choices to make because (once again, Yes) Tucson w as becoming so violent. I watched the environment (literally) "drying up and getting dustier" as developers tracts replaced the (*an exact repeat* I was later to witness in Honolulu) trees, disregard for chains of desert life, it had taken millions of years for nature to slowly adapt to the arid sands which extend down the Sonora desert past Guadalajara. I was getting robbed doing things like walking back and forth to the "Frys" Supermart, only half a mile away. Month's later Neil, my best friend was murdered by his closest neighbor at point-blank gun range. The big cities of the USA have become a very dangerous place to live, *even as* each major city brags how much new money it spends on police, equipment and *distraction from the real issues* facing society's survival. On "the Islands" there's too much commercial push, moneygrab!

Are people going to cooperate... or compete, will governments leaders mature to stop their lies and wars aided by the increasingly small minority of ultra rich and powerful who are witnessed in objective reality as "The Masters of War" and dictate, period!

IT WAS the plight of Bismark that he lived to see the economic rise of the banks and a new class which dumbfounded his sensibility, a class of aspiring rich who bought their way into a society which had to admit that Economic power was THE thing and even began for the first time I might add, to transcend the pre arranged classes of aristocracy and bloodlines. It became more apparent who owns Hawaii when I looked at the list of land owners, especially for the fruit and sugar cane producers.

All THESE forces of change are VERY MUCH, still in flux in Hawaii, the new rich, the OLD POOR, the recent immigrants, and ancient, time honored families of Hawaiians with Royal Blood in their ancestry from only a few generations to today.

Part One
Long Prior to Iniki
Chapter List

From Hawaii to the "Golden State" the Environment becomes a matter of disposable debris, clear cutting and Mother Nature is on the run, waiting the reckoning of all this. The reckoning is Iniki.

1. DAYS OF WARNINGS, END OF AN ERA—A DEPARTURE TO REMEMBER
2. PURE CANE FROM CALIFORNIA AND HAWAII—PUTTING UNDERSTANDING TO SOME THINGS IN COMMON
3. KEEHI LAGOON—INDIVIDUALS DECIDE, WHO MADE A DIFFERENCE
4. SUNDOWN ON THE REDWOODS—HISTORY'S BIGGEST ROBBERY
5. CALIFORNIA CORPORATE BURLESQUE—"CALIFORNICATED" BUT NOT YET IN THE LAND OF ALOHA
6. DIVIDE AND CONQUER THE CITIZENS—"YOU ARE LIVING IN A POLITICAL WORLD.

END OF PART ONE—PRIOR TO INIKI
CHAPTER LIST PART TWO, "THE PIERCING WIND"

7. SOLITARY SCENE AS INIKI BEGINS—"ALONE AGAIN AT HI.
8. STORMY CROSSING OF KUHIO HIGHWAY—"NO ONE HERE GETS OUT ALIVE."
9. ALOHA HOSPITALITY AT A LAUNDROMAT—"YOU MUST BELIEVE TO RECEIVE."
10. SOME INSIGHTS, "THE PIERCING WINDS OF INIKI."—"THINKING HAWAIIAN IN HI."
11. CHANCE MEETING WITH AUSTRALIANS—"GOT A MATCH AUSSIE."
12. SHELTER FROM THE STORM—"ANY PORT DURING A STORM"
13. NEW MORNING ON A NEW SHORELINE—"WHERE AM I?"
14. GETTING GUARDED GAS AT KAPAA
15. CLEANUP TIME ISLANDWIDE BIG MESS, MORE TO COME
16. LEARNING TO FORGET SHOCK!
17. PARTYTIME IN KAPAA
18. CONCLUSION—PLIGHT OF THE ALIENS
19. CONCLUSION—BIG SAVE SAFE SAVE AT HANALEI BAY
20. CONCLUSION—HURRICANE INIKI—ONCE IN A LIFETIME EXPERIENCE
21. SHORT STORY OF HAWAII—SHOOTOUT AT THE HANALEI GOURMET
22. THE FOUNDING OF THE THEOLOGY OF RECONCILIATION WITH THE EARTH OUT OF THE RECKONING OF THE HURRICANE INIKI. THE FOUNDING A ST. CLOUD, MN SCHOOL OF ECOLOGY LITERATURE BY DR. PATRICK A. O'DOUGHERTY, ROLLIE CLARK

AND REV. MARTIN RATH, OSB. OUT OF ST. JOHN'S UNIVERSITY. HAWAII IS TURTLE ISLAND.

PART ONE, LONG PRIOR TO INIKI
CHAPTER ONE
DAYS OF WARNINGS, END OF AN ERA

FOR FOUR days, Hurricane warnings and obvious preparations were taken, I saw progress of all this up and down Kuhio Highway, the most obvious was taping up windows, all over Kapaa. "Di" was gone, less distractions. "DI" AND I had discussed her "departure date" on many occasions and then again it was set for Nov. around ---. . Her irritability constantly bordered on violence, the nerve grinding I could do without. There were recurring and ongoing patterns which I got better at recognizing , this took years because she kept a duality to her personal affairs which I later found incredible but then "Love is blind." When you Love one another , lost in the depths of care, concerns and interests of our mutual welfare. Being realistic, we had been doing well, against incredible odds. "SHE'D HAD THE BEST" of everything, That's how Army Officer Lt. Victor Luna described "rearing" as the only daughter of a General. Victor was assigned ...her personal chauffeur. I certainly wasn't equipped to do that. It was fortunate she got picky when she did before our camp flew away, she did! The discipline thing was an area that "DI" was rebelling about from the time I met her, so to lay down even more of the same was a fine line for me to tread. The departure date from Lihue (Kauai'is main city) to San Fran. Was confirmed, set and she had her "Ticket to ride" back to "MENDOLAND" on the mainland, home of civility, Californication, Wine and hot running water. She missed that more than anything else she ever discussed, but her ......!

"PAY FOR YOUR TICKET and don't complain" I'd told her more than half a dozen times. The "intentional redundancy" child-psychology seemed to work. Towards the very end I asked, then told her to move her tent-campsite about ten yards closer to the oceans roar. Just so far and just so close, yet I could keep a safe eye on her.... not above the Pacific's roaring surf and fresh air. I was trying to be as fair, evenhanded and at the same time practical as could be delt under the circumstances of our many challenges with various bureaucracies. We were being jacked around with things like simple, basic housing by "authorities" posing as friendly workers from their "Dilbert Cubicles" in Lihue, the way they work is, delay!

WE WERE then on the East shoreline of Kapaa near town center with her favorite HI. coffee and coffeeshop nearby "The Decko-Gekko" that place was just shutting down permanently. The new "waihines" there were a jealous lot. She'd maintained that punctual ritual for days, months and all the years I'd known her, she had to have that coffee, morning and afternoon maintained with a ritual of imitation art-study, the same stuff over and over. That was awful... or so others told me. These were constants in her "Roller-Coaster reality" I'd slowly learned to accept over the years of our engagement   THE DATE Sept.. 6TH. A beautiful afternoon in Kapaa as Di, a friend Jeremy and I walked out to Kuhio Hwy., to see her leave on the bus for the airport in Lihue. Jeremy was going on up to the mountain visiting friends in brilliantly green -florescent foliage near the edge and end of civility. That's where "Princess Di" left me, on a whole lotta levels and I, in turn was stepping out for a visit to the "BIG SLAVE" store for some Stunlager beer, A cure for excess reality and medium butt pains.. This area is just a few miles South., where the Tarzan Movies were made with the original Buster Crabbe as Tarzan. I could right then identify with that scene a lot and even " Moh Betta" after she was back on the bus.

I FELT I'd seen better, "Betta" days, next morn and was glad to be all done with babysitting in the jungle. Within X no. of hours, the first report of a Major storm was on the radio... just hours after she'd gone. What phenomenal timing for the fury of Mother Nature! That thought crossed my mind a lot. We'd shared natures very finest here and there, camping and roaming the Northern half of Mendocino Co. What we witnessed in its essence was more, much more than the finale to the cutting of the last of 97% of the ancient Redwoods, It was an era-end of the "WINE AND CHEESE" mentality for Mendocino and CA.

What more does California have to do with KAUAI'I and a hurricane? Plenty! There are many parallel developments in use and abuse of natural resources. The HI. Elder I spoke to in Kapaa stated that HI. People are being corrupted and in turn Nature will again (She said this after the hurricane) bring down natural disaster.

CHAPTER TWO
C. AND H., PURE CANE FROM CA. & HAWAII

Within my own memory of Hawaii, both Ca. and HI had been much more pristine and above all else, California and Hawaii lands were more accessible, open for the use of all people, only a generation earlier to those who wanted to explore the beauty of these unusually magnificent areas, certainly the finest of nature I've seen in more than 155 trips (in terms of added-total mileage) around Earth. Island-dwellers hope the beauty and purity of the Kauai'I environment may remain and not become "CALIFORNICATED." Previous to the crucifixion invasion of missionaries and capitalists, were hundreds of years of other power struggles. War isn't new to the HI. people and went on for hundreds of years. They endured various warring chiefs, hundreds of taboos, infanticide, superstitions and fears of the unknown beyond the reefs, mind and body boggling taboos. Then came the successive and successful "Howle" invaders of their culture, "christians" and capitalists joined forces with the USA military to depose Queen Emma in Honolulu at her palace. This was the major turning point in modern HI. History It could be said factually that Kauai'I was never conquered. This is a very significant thing to remember when you read about the people, an isolated group of people... on Kauai'I today. The majority has never left their island and as of a survey less than 20 yr. ago, the majority today never wants to leave it, period. That says more than earlier pages could ever relate about the spirit of Kauai'is peoples. They are very special and indeed consider themselves so.

THE U.S.N. recently moved its main submarine fleet from WA. to the West Coast, BARKING SANDS was the scene of Hawaii's ancient warrior ceremonials, cliff jumping for battle losers. The irony of this is the very ancient practices with the right now, modern. As a further contrast with Kauai, the other end of the isle chain is the "Big Island" which was conquered and united under the rule of King Kamehameha the Great. They were also hit with white-mans diseases. Over 80,000 Hawaiians lived there in the early 1800's until a variety of diseases from "Island Visitors" did their thing.

JACK G. LONDON, California's greatest writer sailed to Honolulu aboard his sailboat, the "Snark" and then worked on various HI. Islands. He was honored to visit with the Queen and Hawaiian Royalty THEN DID HIS WRITING; accurately described these East USA coast christian capitalists and land grabbers equipped with visionary levels of greed as the following (from over a hundred years past) "Here in Hawaii it's the Churches TOPSy TURVy LEADERS, sitting at the Head of the HI. Tables!" They have taken and taken , then stolen the land itself as they enjoyed from the far West, Kauai on East to the Big Island, the most of "Paradise" remaining on the planet. Yet in a spirit of material pursuit, These whom the King fully expected to take their wooden houses back home, have gone farther and farther away from basic respect for the native people and the land itself. This was part of the churches message that thousands witnessed after Hurricane Iniki , their arrogance was unmistakable. In the absence of Brotherhood, simple respect for the peoples , the lands and life forms, those elements which Iniki unchained in nature, will in time again return with great loss of property and next time , human life too! This is in essence what the Hawaiian elder told us after the hurriicane. California has seen a lot of this change of heart, during the thirty plus years I lived in and out from So. CAL. All the way up to Humboldt County, the No. border with Oregon. The parallels between Ca. and Hi. are numerous, unmistakable and rather depressing to people who've lived and visited both, indeed!

I spent a lot of time sailing around Keehi Lagoon and watching the "Weakend Vikings, Howle (foreigner) Tourists and the careless piling, a tide of trash , some of it going back to... would you believe....WORLD WAR TWO. At the low tide, far end of Keehi Lagoon I worked with many friends who recycled metals , (East of HNL-Airport) friends like Billy Strong, Jim, Dave, Justin, Keeki, who cooperatively "fished up and out" tons of plastic, metal, wood, boat-wreck parts, toxic wildlife killing trash to be recycled. This was independent free-enterprise business in its essential, simple, human form and for us workers a modest income. And too conservation at its essential ground floor beginning! "SECURITY" saw us as a nuisance. We worked hard, played and had cold beer later. This was my "Intro" home in Hawaii, Keehi Lagoon, and an excellent cross-section of the best, worst and the rest of the HI. Fabric of life which centered around the Pacific, here in the center of the Pacific, its crossroads.

CHAPTER THREE
KEEHI LAGOON

KEEHI LAGOON was a neighborhood and a community of diverse peoples living on their boats, BIG YACHTS, catamarans, trimarans, cabin-cruisers, AND SOME Christian Scientologists on an old Navy barge, all the above with hundreds of world visitors "tied up" while they discovered the "Aloha Spirit" of a time-honored anchorage dating back to Hawaiian Royalty and Hawaiian Homestead lands Now it's just hurricane-wired fences with "beings" behind cubicles who don't even know one another. We were a group of hundreds of friends who lived, worked, and celebrated life, most of us on a small but honest income. The State and the "Fed" wanted change, because we were INDIVIDUALS who really made a noticeable difference. The actual content of the waters in Keehi Lagoon should have been concern NO. ONE. From an environmental standpoint, this area should be kept very clean because its central to Oahu, location as a central fish-breeding grounds for that heavily used and fished area of Oahu.. After we were harassed out, the "Feds" took over, they ruled and chickenshit-reg'd us and hundreds of other living on our boats, making a modest living in a spirit of joy.

THE NEW FINANCIERS, nervous and well dressed Japanese Samurai of the Yen and Big Bang Buck appeared at the front door of the Lagoon General store, a place where we'd formerly walked in and out marking the first coffee of the morning and the last beers of a hot Honolulu's days work. Then even the State of Hawaii lost their control and interest in Keehi to the Fed. Govnt., the rest is history yet not one dam word was in the local Hon. Advertiser news. It's interesting , fascinating to witness this degree of terrific teamwork. Funny how even the State pigs missed the trough that time!

Here I mention a few of some noteworthy parallels between HAWAII AND CA., in matters of the environment or rather, lack of concern for it. The situation at Keehi Lagoon was an ecological disaster which our Military and industry had certainly encouraged and "built up" with the aid of thousands of private boatbuilders who used these waters for a sunken and floating dumpsite after working on their hulls, picniking, partying and the daily going-ons for generations. The "Heavy Metal" debris stuff didn't arrive until mainly after W.W.II but as recycling efforts indicated over the next year of diving the bottom, there was plenty of "sources" from all sectors and some that not even a team of local metallurgists could identify over at Kojimas, metal. This kind of waste spells death for fish trying to breed.

CONVERSATION about CONSERVATION, any conservation wasn't taken seriously in Hawaii for generations and by the majority, it still isn't today! For those who doubt the validity of this statement and observation, they really ought to go out and get in contact with the majority of the people who live there at any given time-period, then the enlightened doubters can return and we'll compare a few thousand observations. The same pattern was true in CA. But the environmentalists did get voiciferous and organized after Conservation became "IN" ...then along came the late 80's and "90's" and it once again was discarded as a popular cause (by the big "IN" conservation-money groups, Sierra, etc who in practice neglected their professed activities) as unexciting, unrewarding UNPROFITABLE. The crossover in mentality between CA. and HI. Is what I'm elaborating here, to relate the mindsets behind public change of a gargantuan California economic power and accompanying psyche to a much-less, mind less plastic money-machine.

Having stated the above, don't think the controlling interests of "C. and H." are short of their sight and mark to exploit what they have. Many indeed the "Ruler estates" of Europe are mirror images of what C&H. evolved into but... in this, "the new world odor", it took about two hundred years. That's progress for you.

"What happens first in California, will happen later in the rest of the USA."
- Anon. -

Mendocino village is an unincorporated hamlet in the heart of where once stood 4,000,000 acres of THE giant Redwoods, and is as good a place as many to begin this crossover exam of Hawaii/California ideas, influences and corporate motivations of the present. Even as rental land, storefronts, public access areas were jacked up in rents, user fees, and most of all "accessibility to the public" was being fenced off, consistently the finest small and medium "public serving" family-businesses were run out as were the poor and perhaps most shamelessly disgusting of all; a new generation of its own youth, harassed out of their homeland.

You have to look at the worldwide pattern to see this correctly, long lines of refugees from civil wars, coming here to the USA. I personally witnessed from 1965 to 1995 from top to bottom of CA. "Mendoland" was changed in various ways as an entire generation was driven out, it became favored for "plastic-moneyed" by the Bay-Area Bed and Breakfast, Weekend-Vikings. To be looking at this as a direction of the society is a matter with far greater implications for HI. The rest of the USA, is watching what happened in CA. , the "Wilson" right have left eight years of their enduring ruthless imprint. The Mex. Americans soon to be the Ca. majority were hit especially hard, denied even basic medical care. A ruthless mismanagement of resources, human and natures. After working and living in CA. for more than thirty years, I finally gave up seeking a better way of life there, it's too "Californicated" in CA. which is what the "Powers that Be" had in mind. Simple enough!

CHAPTER FOUR
SUNDOWN ON THE REDWOODS

INTEREST AND PUBLICITY funds for "Show and Tell" educating the public to save the few (in comparison with millions of acres more, an earlier generation) surviving ancient Redwoods at Carlotta, had virtually ceased as of 199- At the main publicity office in Ft. Bragg, I know because I staffed it. Coincidentally, there is the Laser-measuring ultra efficient state of the art sawmill where more board-feet of lumber, specifically Ca. Redwood, were cut than any other mill ever had or likely will cut, due to the very gigantic size of these ancient giants. Also during this period, the library in "FT.DRAGG" (As the teens in Mendocino referred to it) was effectively almost totally closed for "Lack of public interest" as much or more than sufficient funds. This has to be the most transparent, arrogant deception-frauds (against the children of California) I ever encountered. It ranks on a scale of lies with the Texas-Oklahoma Savings and Loan Fraud theft of Billions Likely even more was taken behind yet another curtain, the "ELECTRONIC-CURTAIN" of their time honored corrupt politicians who hushed numerous bank collapses as did the F.I.B. (?) who had cooperation with the nations press, maintaining a "REDWOOD CURTAIN" OF MEDIA SILENCE WHICH IN TURN ENABLED THE CROOKS, AKA. THE MILLIKEN CROWD to pull off their frauds with taxpayer money backup. That is, taxpayers paid the tab once and then twice, to the total of approx. 3% OF THE USA G.N.P., IN ... I BELIEVE IT WAS 1996 THERABOUTS.

"SOME PEOPLE ROB YOU WITH A SIX-GUN AND SOME WITH A (FOUNTAIN) PEN."
- WOODIE GUTHRIE -

DURING this "END" period, immediately prior to the Clinton Adm. obfuscated, fogged over and well couched "Sellout" (They described as a BUYOUT) of the Ancient Headwaters area of the Redwood forest, much was accomplished from "Behind the Redwood Curtain." I personally witnessed dozens of the most dedicated , hardest working and totally volunteer conservationists being slowly but surely intimidated and harassed into myriad forms of submission. The way an ex-Vietnam vet put it, "Their policy is one of divide and conquer." Equally scary is the other side of the coin; How many from within the Govnt. were actually engaged in this corporate sellout? The nationally circulated FBI posters for robbery of the TX. And OK. Savings and Loans were a clue that few, one in ten thousand may have caught on to.

These crooks pulled off a robbery that dwarfed even the gold-brick robbery of the Third Reich. Try adding up $50,000 per tree, times over FOUR million acres of Redwoods! IS THIS then to be the plight of Kauai , will its trees be clipped down to the bare sands as we saw in Mendocino, CA. Kauai's soil has already been ruined by saturation with chemical fertilizers with over cultivation of sugarcane by C. and H., Mc Bride and other corporations! The local Kauai doctors REFUSE local Kauai water, they get the bottled stuff from Lihue. Parallels arise and vividly I recall going to Volunteer secretary work at the Ft. Bragg, Ca. Redwoods office, a group hdqtr. for local conservationists. No less than a Bomb scare had been called in only hours earlier. It entered my mind, I might be dying soon in that office, as I cleaned up the desk surface and carried out some trash to the sandy soil outside the rear door. There were then current activities by ANTI-CONSERVATION ELEMENTS in many states including AZ., CA., ORE., WI., MN., I WON'T GO INTO THE FBI END OF IT, SUFFICE TO SAY in CA. JUDI BERI was still actively speaking to save the ancient Redwoods as her end neared, victim of a car bomb.

"It's Sundown on the Union, and we're living in the USA, it was sure a good idea until the trees got in the way." B.D.

Northern California in the late "80s" was the home of many Ex-Bay areas "BEST AND BRIGHTEST" who'd slowly been beaten back , steam-rolled flat by commercial interests and big money. I didn't get to live there much until about 1990 and regret not being around to witness and work to save more of the Redwoods and then it would (in retrospect) have likely done little to "nudge" these corporate powers. Specifically in Mendocino Co., I recall getting to know, witness a very eccentric judge in action, whom might have made a lot of difference over a time period. Once again, it's impossible to predict what one person can inspire or do. One things sure, the corporate security system is piling up some mighty high and expensive walls even as the governed get distanced from their controllers.

All this with the Redwoods (it didn't take an Inspector) was mainly attributable to long established logging interests, armed with big bang money from Wall Street's then accepted farces, practices including , paper shuffling junk bond dealers arming their portfolios from shady sources with bottomless pockets. They then took all this to force buy-out of P.A.L.C.O. (THE PACIFIC LOGGING. CORP.) which also entitled the bearer to the ancient Redwood lands via money from junk bond stocks sold by Milliken. I'd worked until there was little more I could accomplish then... witnessed monthly increases in various forms of harassment (thievery) as "security".increased!

"A society which is just and tolerant, is in the long run... secure."
D. Halberstam, correspondent in Vietnam and likely
one of the greatest living Am. Writers.

CHAPTER FIVE

CALIFORNIA CORPORATE BURLESQUE
"THE BIGGER THE LIE, MORE LIKELY THEY BUY." Anon.

There was violence and I'd been hospitalized from "UNKNOWN" forms of injury several times in Mendocino, Co. after instances too numerous to recall of "security harassment" Time after time in HI., CA. It's noteworthy to recall the final trip I made to Carlotta, Ca., for the final demo. for the ancient redwoods there. I clearly recall walking out and back towards the main road, then the next thing I knew, I was lying down on the concrete and being awoken by police officers, it seems someone had "possibly hit me with a board" and then the police said it would be better if I took a ride (at taxpayers expense) to a hospital for whatever. Again and again I have witnessed the final product, hundreds of various agencies that we pay with taxes, who simply are without conscience and grossly intimidating. Then there are the Government and corporate POLLUTERS WHO STILL DIDN'T BOTHER TO RECYCLE anything, as of the mid- 90's.

To put things in perspective from a far different source (another authors quote) "It's no secret that the secret is out, the American Century is over". "The reckoning" for all this vertically integrated bureaucratic B.S. , draws nearer even as resources are carelessly flipped to the four winds of capitalism. Natural laws and life ignored in the course of the fast-track buck, a disease reaching new parallel heights of absurdity between a quartet of strange bedfellows, the Military, Politicians, Right wing religious and of course the Multinational-corps who big wheel deal with a well lubricated global machine Corporation, headquartered for the central Pacific... in the City and County of Honolulu, USA!

What has the above to do with Kauai, it's has been the most protected isle because, so far the climate and people were withstanding the "riffraff" but I personally believe that Hurricane Iniki changed that just as the storm took down a lot of the "Deadwood" from earlier periods, all over the isle.

PERHAPS the most obviously outlandish example to mind is the Environment. Protection Agency of Hon., who'd placed half a ton of misc. everything from drafting tables to dumpsters of office paper just behind their massive offices , all waiting to be buried on the side of a mountain just a few miles past the Wilson Tunnel.

A LOCAL to Lihue, Kauai individual told me, "I can see the time coming... the logging of Hawaii's Garden Isle will go on." When it does happen it will likely be 24 hrs. per day as I witnessed it in Ft. Bragg, Ca. This isn't actually inevitable yet very likely because of the current climate for business and the conditioning of the USA and HI. People. Big business is getting better at total-collaboration with all levels of govnt. and this is due in part to apathy but also select use of the media and computer efficiency.

As has been predicted, the Hawaiian people will likely stand divided and in a semi-idle position when this does occur. Why predict this? Because these people have already witnessed and waited more than a hundred years, doing very little to change an illegal deposition of their Queen. Even as the last of their lineage and percentage of their people are steadily declining, the "powers that be" are indeed observing all this and
continue their "work." which is easier in the absence of the indigenous peoples.
Another way of viewing this ongoing process of elimination, is simply to "Leave less witnesses" AND disable the reporters of /for any independent observations.

THIS "process of elimination" is steadily occurring with the Natives of HI. and will likely be done as we saw it in Ca., "CALIFORNICATION" The native sons and daughters were literally driven out of Mendocino and Humboldt Co. by the commercial people, harassment, denial of shelter (affordable housing is now but slowly became non-existent. The structure was turned from a low-key colony of artists, musicians and writers, to one of an ultra-commercial, bed and board-business hype which finally boiled down to an ability to make a living for thousands, driven out of that whole part of the state. It was all pretty damn gross, the commercialism which simultaneously on various levels of business resulted in the cutting of the trees and worse, the loss of a very thin topsoil which may take thousands of years to re-make,. Put differently, there once was one 200,000,000 acre Redwoods, none other on Earth.

A LARGER pattern I unmistakably witnessed on all the HI. Isles I visited (from approx. 8-89 to 9-91) is occurring globally and points for the thousandth time to increasingly unrestrained power, uncontrolled save by privately delt negotiation, sweetheart deals between Multi-Nat. Govnt. And family combines, Mc Bride on Kauai and Cargill on the mainland are good parallels. They don't do business, they dictate what they will do next to the planet with little conscience beyond PROFIT. MULTINATIONALS, like LOUSIANA-PACIFIC , the worlds largest private landholder and Mr. Junk-bond Millikens "Palco" know who they are and who you are not to them! Average citizens/working people don't count!   I want to remind the reader that these various CORPORATIONS are controlled by individuals operating on Kauai, farthest West to Manhatten, farthest East USA in a computer age where finances are leaving an electronic trail, mainly invisible to the eye, many times there isn't even that, with multi-million dollar trades and stocks.

CHAPTER SIX
DIVIDE AND CONQUER THE PEOPLE

One, two then three generations in our family experienced the same thing I did, to earn a living it was necessary to leave the family and go work for the "giants" in the city. The control of environment and destruction of environment is the long run result of our growing mainly city based, human race. This leaves the rich to buy up, control , own more of the land in the country and isn't my choice to do anything better, this pattern is occurring on all the HI. Islands and was somewhat accelerated after the storm   Multinationals have removed thousands of "UNNECESSARY" workers under claims of prosperity, after committing "UNCLE SAM" to billions in corporate welfare, even as the general population remained "I don't care". .. in a lukewarm reception of this news, most didn't learn or bother to.

Why, for that matter will the same pattern continue to expand in Hawaii as it has in the past and lead to further cuttings of her unique forests. Because for far too long the indigenous home-landed, native peoples have been cast and broken asunder, worldwide IN THE NAME OF PEACE, PROGRESS AND PROSPERITY for a few corp. giant families and hundreds of conscience devoid Multi-National Corporations. THE FAMILY unit continues towards further isolation, parents separated from their infants and tiny children, in Daycare centers. This practice in the long run is the recipe for social disaster , anywhere! It constitutes a net effect of "Divide and Conquer " for the foundation of society, the family unit. It was pretty obvious in its effects when I visited a gathering of the last Hawaiian Royalty at the Queens Palace and later met the woman who would be the next-of-line Queen of Hawaii.. By standards of International law, she should be queen of HI., now! Here as worldwide, the indigenous peoples are being grossly manipulated by the same corporations that had them a hundred years ago, not much new!

"You are living in a political world"   Anon.   My very fair offer to donate money, writing time and energy towards " Hawaiian Unification" work. was largely ignored and this happened during a year of peak activity by the loyal native, indigenous peoples, 1997. Let me say this, I was pretty damn disappointed with "their treatment " or rather ... lack of commitment and or response from letters sent to (unspecified) various leaders on Oahu, the Big Island, and Maui! In spite of many thousands of hours of writing and researching and a lot of running around after writing hundreds of pages with conscientious concern expressed to the various leaders, no responses! Then later I had numerous contacts with then Mayor J. A. Y., of Lihue, she was so busy after "Iniki", it's really difficult to comment about her, did she do more good or what ever? I did some correspondence with the newly elected mayor of Honolulu, Mr. Harris, as expected, I was ignored yet it's still for public record best to speak ones conscience, especially when things are really rotten, if you survive! Responses are usually measured in degree of "sidestepping" issues when dealing with politicians.

CHAPTER SIX
DIVIDE AND CONQUOR THE PEOPLE

I can still feel strong about all this effort, it was work that was much needed, regardless of the political winds and stockmarket shysters.. . and furthermore as contradictory as her words, letters and actions remain as a matter of Public record, the Mayor of Lihue seemed to indicate she valued my work, legitimate concern for protecting the environment from further solid waste pollution, etc. She was doing some fine watercolor work in her office and couldn't be reached for further comment at that time. She had finesse in her art, I watched her carry ... some of it out the back door of her office in the county bldg.

The thing about care of the environment, it's not a real politically glorious subject and here on these islands, you can't take it to the bank! The people who've been here for generations speak with the greatest conviction; Hawaiian politics are rotten to their not so ancient core!

I had some interesting contacts with politics on nearly every isle, it's the same old game, just a different head nodded in a different political breeze with screwy local rules, still different on the isle just miles next door a mere twenty-five minute flight. There are different rules, more games, runarounds ruled by a mean machine few natives know.! George Anderson of the NO. shore explained some of this to me at the Court House gathering of politicians in Hanalei Bay.

"I am running for office on this (KAUAI'I) island but have to go through the political machinery, the traveling expense and time and forms and red-tape ... on Maui."

It doesn't take a genius to figure out why the control of election of honest officials is for the rich and those contained within the party machinery.. I vividly recall the election process in Ca. And going through an election-ballot form that was difficult enough for a college grad., to decipher. Here in Hawaii, the median educational level is (approx.) between sixth and eighth grade. Professionals who have worked here an entire lifetime have said the college standards are significantly lower than mainland.

PART TWO, THE PIERCING WIND
CHAPTER SEVEN
SOLITARY SCENE AT THE CAMP
INIKI BEGINS APPROX. 4:45 P.M., SEPTEMBER 11, 1992

I WAS calm, confident looking forward to an entirely new adventure having some... basic ideas about a hurricane, things like some fast wind and driving rain. This wasn't going to be just a new adventure, This was to be a major storm which not even the Natives comprehended. Other than getting some work done then going swimming around noon, I was expecting nothing earth-shaking, especially during this period following the separation with my partner. If anything, I was more relieved than in months. Living in the Kauai community is daily an experience in Peace and tranquility.

ALTHOUGH "DI" had gone, there were just too many recent recollections and reflections from the whole experience of living and being together, sentiments complicating, cluttering my mind that soon would be just plain... overwhelmed. I was and am now a guy who always preferred "outdoor living" to "social suck-cess". Jeremy the Oregonian walked by carrying his giant green backpack, he smiled and exclaimed "Did you notice that the wind has calmed?" "No," I glanced around and the tree leaves were limp. "This is the calm before the storm and I'm going to a Hurricane-partying up in the hills." The islanders love to party. It just so happened I didn't choose to be with them, to me this storm was a matter of interest. I wanted to be outside and witness the spectacle. I'd never seen a hurricane, so "ANYTHING IS POSSIBLE", as they print on T. & C. surfgear, wavetools, etc. I had that tent placed, carefully behind the most monstrous Ironwood and was confident in survival.

THE DAILY routine of taking a walk, going visiting and picking up my survival supplies was combined with a ruthless efficiency, I began by making out a shopping list on my left palm then winding on out Mr. Eds horse path, past the Hi. Gravesites, ocean shoreline and then on in farther to convenient "civility", approaching our very own carbon-monoxide belt which indicated true North-South bearings with "Slaveway" store in sight. No matter how wild things in the night might get with visitors or "DI", I could count on this normalcy. This "emergency supply" visit was within minutes of the storm yet people seemed calmer than say, at Waikiki on a Saturday. That gives an example, in perspective of Kauai vs. Oahu, it's a slower, more humane aura, not this money-push runaround hype of a show business nature.

NOTABLE (a-hem) of hurricane "preps" on my shop list that day was the Hawaiian RUM. I entered the jammed market, got three items and noticed the line of shoppers extended to the rear and back to the far left (kitty-corner) corner, on the opposite side of the S-mart. People had diapers, candles, matches, candy, beans, fruit, soda, beer, canned this and that. When one of the HI. women saw my three items, she said "Go ahead" and I remarked at this gentility and gracious spirit of these peoples, all this on top of the threat of an imminent hurricane. Would this happen in Texas or Spain, could it? Would people be so considerate and humane? I really doubt that it would occur anywhere else but right here in Hawaii. These people have their priorities down. They are basic and have some incredible instincts,

CHAPTER SEVEN
SOLITARY SCENE AS INIKI BEGINS

a capacity for sincerity in their smallest of affairs... which is likely why the Politicians and Businessmen have (paradoxically) been so successful in cheating them! Yet there is a silent dignity and greatness I felt in their presence, not only the short visit I had with the woman who would be the next reigning Queen of HI. She had a "Power of Silence" and "Presence" more than anyone I ever came in contact with in my entire life.

"THE USUAL" group of Surfers, tourists, beachcombers, noontime beer-drinker non-comprehenders, expenders, pubic pretenders at the beach and nearby playing field, next to the Kapaa Pub. Library were solid-gone like the earlier afternoon had disappeared and I never saw them again. Guess they were evacuated out with their airline of our choice, the U.S.ARMY who were briefly operating, for evacuation purposes with the C-5 A, the worlds largest air-transport craft. I wondered if the airstrip could take (the weight) it.

AMONG the many friends I never saw again or waited about a year for the dust etc. to clear were RICK the sailor and guitar player, CAROLINE, BRIAN FOSTER, WAIKIKI WILLY, KNUTE, RON, DIANE , GARY, BIG B. JOHN, Pacololo JOE, JIM and his Irish Uncle Mike the pilot and too, a long string of other "accquaintances" and smoked out phonies . Many I wanted to forget anyway, Amateur- Bogarts, violent con-artists, and more who arrived during the P.I.S.S. segment of island recovery. SUFFICE TO MENTION also the Politicians who generally traveled with a team of other liars, more liars, petty hustlers and third rate capitalist sellers of what have you. These "low-life" were around and ninety plus percent just disappeared when Iniki hit. They didn't have the staying power to come back. Still many others later collected such a pile of insurance money they left town and likely visited on the mainland. The Lord works in mysterious way.

CHAPTER EIGHT
STORMY CROSSING OF KUHIO HIGHWAY
"NO ONE HERE GETS OUT ALIVE"

I'D comfortably situated myself in the tent with all... of few "INIKI" preparations, learned from many lesser storms on the Atlantic and the Inland Sea of Lake Superior. Most of what there was to be learned about storm survival was handed to me from sailors I'd worked with and met in British Columbia, Louisiana, Minnesota and some of the Saltys from W.W.II Talking to sailors who'd weathered Hurricane force storms, I got advice and waited to use it, which consisted of the following, try to have the Basic supplies for successful survival, situate yourself as safely as possible, relax and pay attention. MAINLY RELAX.. O.K., SO FAR! The most vital "prep" was moving the tent back from my desk which was firmly nailed to the So. Side of a tree. I packed up and had moved the tent back, around four days earlier, behind the largest tree I'd seen on that part of the isle. Someway I found out it was a Ironwood, even the name suggested strength, but really, this was an exceptionally massive tree, later after the storm had uprooted it, I witnessed the taproot system, spread out for about twenty-thirty yards in all directions. Because the tree had been rooted in sand, the

CHAPTER EIGHT, CONT.

whole root system just peeled up with the winds force, like you'd peel a skin of a banana.
It was usual to do some writing every day, that day I did none.

I was staying tuned into the hurricane updates and then they said the hurricane was coming, maybe about an hour because it had made a change then it had started to rain, it got heavier. The sun was shining and then it all got Grey. Suddenly I got a feeling to get up and leave, go, there was no rational explanation for leaving. I'd been "In" for less than an hour. ABOUT THEN (GUESSING BETWEEN 4:30 AND 5:30) I DECIDED TO GO noticing the wind was picking up. The Sony radio was off, I knew not the latest update on the storm. Who would say they really cared after days and days of the same weather forecast. Since Jeremy had told me "Iniki" was overdue. Factually what had happened, the storm had just touched the SO. shore of Oahu, headed on out to sea, then it made a major turn and came up towards Kauai. This was the end of the calm , prior to the real fury of Hurricane Iniki.

A STRONG FEELING to get out then, was all I had to go on. I really scrambled around, thinking "What am I going to go with and whereto". The tent, it was a small affair and a tight squeeze with only the blue Berkeley Co-Operative Bac-pac. It was a day and a half size. Little of real consequence was inside, I.D, a few clothes, some money, pens, socks, etc. I was getting instinctive and swift about my motion and decisions. Distinctly, this was life or death, I realized!

UNZIPPING the tent flap then rising out the door and rain cover, I noticed the wind was stronger, much stronger than I believed possible. Remember, it had been a dead calm for about forty-five minutes or so. It was getting noisy, whistling but nothing worse than a heavy rain storm with wind.

GETTING OUT of the camping area was a challenge because tree limbs were starting to fall as debris was loosened , leaves were being peeled off the trees. The jungle layered canopy of fine vines, moss, was starting to shred and rip up. Wrappers, plastic bags, supermart stuff was swirling around and it didn't bother me a bit. Yet the wind was getting stronger by the second as I circled my way on out about an eighth of a mile in a semicircle to the right, then straight down the horse trail of "Mr. Eds". There wasn't going to be my usual noontime beer and swim today.

The pink and green bikini "Di" left behind was now flopping around on a post marker just before the path turned West. I glanced at it and noticed it was now a wind vane with a nearly vertical level.

NEXT TURN to the right, after jumping over and around dozens of broken and rotting branches and tree limbs in my path, led within sight of the driveway, then out to Kuhio Hwy. This driveway was never actually paved or graded, it was a trail for Mr. Ed. and hundreds of tourists who regularly went to the beach. The feeling of living in this area, camping on this field of trees and flowers was a special gift. Rumor was some eccentric lady owned it and wanted it, like so! Some phony fly-in realtor from the Big Isle and other archaeologists were the only other "officials" ever here.

CHAPTER EIGHT
STORMY CROSSING CONT.

SO, I didn't look back. I felt very "antsy" and kept going at a fast trot, that final forty, fifty yards through the furrows, chuckholes until I could see Mr. Eds water barrel was gone. Had the wind taken this away? Still looking in that direction, there was no one to see! Usually there was heavy vehicle traffic, dozens and sometimes a hundred people walking and wandering around from the near sidewalk right up to the far side of the massive parking lot at "Slaveway" but now just a bunch of airborne debris of nameless origin and unknown destinations. "Gone With The Wind" would have been (MY OWN TRUE LOVE) a splendid background song. She was gone and everything else was loosening up.

IN the distance behind me down Kuhio Highway only thirty yards was where I'd met the two big Australians only minutes earlier, and I never found out anything more about them, gone, gone with the winds of Iniki.

There was a curved visual rise, an illusion I'd witnessed before at sea, caused by THE RAIN AND THE WIND IN UNISON, in the road, the wind had created a visual illusion of a ripple in the pavement, the rain was stinging like it was sandblasting me, that sure could have been. I had one last glance to my right and can't recall if the massive tan Ironwood tree, that I'd placed my tent behind for protection, was still standing. Likely it was not, That's jumping a bit ahead of this narration and story.

THERE WAS nothing left to do but start again searching for really strong, stable shelter. I like water, swimming and the rain so really, my mind was still a pix. of calm, unlike outdoors. This outdoor picture was getting pretty outrageous, worse by the moment. There was no trouble walking when I was still in the field just minutes ago, now it took a lot of strength to balance at all and control of direction was getting much more difficult. I can only recall one other example of this force, power of wind, that was coming down into Spain across the French Border on the Pyranee-Mediteranean coast, the wind then "rolled me down the road" bruised but I didn't get hurt too bad. On that particular occasion I recall a small sailboat blown off the edge of a Pyranee Mt.

I didn't realize at that point in time, a large group of people, a search-party had been out looking for beach visitors, etc. This I still find difficult to believe, yet ...this was the procedure. The beaches and my camp area were being checked for tourists or just anyone left outside. YOU THINK OF THE MANY REAL POSSIBILITIES FOR NEEDLESS DEATH AND SUFFERING WITH THESE KIND OF NINCOMPOOPS IN CHARGE, VERY SOBERING INDEED! The point of diminishing returns, that's something to ponder as you tell yourself, the bigger the system, the more vulnerable it is to assorted maladies.

CHAPTER EIGHT
STORMY CROSS -THOUGHTS, CONCLUDED

THAT was the beginning of the beginning of more asinine goof-ups these emergency nincompoops were volunteering, some paid too, to do. Later came the intentional actions and intent to harass and quite simply stated, NOT HELP, give little or even a very poor grade of help to hundreds of the least able and most in need. Let me ask the reader, What emergency-Worker should be given this responsibility IF THEY COULDN'T SPOT A FLOURESCENT (BRILLIANT) YELLOW PUP- TENT -FOR –TWO, within a matter of Yards where they had to walk by. Keep in mind the tent was in direct sight of the beach sand, the Pacific oceans water and a four minute walk to (that part of) the main supermarket!

FOR YOU readers waiting to hear progress about the storm from official "sources", I did some statistical, checked with govnt. Publications, other reports, radio reports, unofficial reports, some accurate personal reports and lastly B.S. from some our very own unnamed national monster you know well. I want to mention the duplicity of this agency's reportage. The report I heard while still in Hawaii and on Kauai was "winds of 200-225 plus, then the wind gage broke", but when I read the govnt. Report at the U. main library , it was toned way down to approximately one hundred MPH less! Well, that's progress for you!

Winds were reported (approx. time) in Lihue, at 114 mph, in Malcahuena it was reported 121-143 MPH. It's good to think of all these reports as impossibly inaccurate because of the three tornadoes which were within the hurricane. This was perhaps the single most incredible addition to the power of Iniki and the one which likely led to its sharp turn in our direction, it had been heading out to sea, then took a sharp turn North, then hit Kauai.HI. does something like this.wind force get "didgitized" or numbered into statistics like most of all we actually have to live with everyday, it simply doesn't and cannot be done!

The US Navy supposedly reported in, up until... the moment their wind-gage broke, that was a 223 MPH, plus wind, likely from one of numerous, possible spots within the tornado swept area which was supposedly near the center of the isle. But certainly in the infinite range and wisdom of the military-industrial grade of coffee or something to that effect. Maybe.

Fair to mention is the incredible contrast in (Hawaiians helping Hawaiians) attitude of the Nat. Guard people who worked so much with any and all displaced people from the storm. I couldn't get much (help, any, anything) but I met some of the Mainland guys from Milwaukee, WI. and really surprised, overjoyed them with as much Miller beer as I could get. Most of these workers were happy to help anyway they could, this was truly enough to bring a hurricane survivor man to tears of joy, it did to me! We survivors got the distinct feeling, these guys were here to do anything they could to help and it felt more like "One of the family".

PART TWO
CHAPTER NINE
LAUNDROMAT ALOHA HOSPITALITY

PLENTY OF POWER, from a Hurricane but no electricity, no heat, drinking water, no safe refuge, but HAWAIIAN HOSPITALITY prevailed when I arrived, stepping across downed power lines. I took a sharp, quick look before crossing this tangle of wire, with that much presence of mind. Inevitably there were to be many more complications later on but for right then I felt a confidence, somewhat borne of ignorance and a stubborn but used attitude I had ever since landing in Hawaii; Somehow things were going to work out for the better!

OF ALL the places least likely to have any sign of life, a laundromat in a hurricane. I could have done a lot better logically but it turned out great, except for the shelter part. After I made it up Kuhio Highway past the bridge, past McDonalds and across the storm blown open area and didn't get hit by anything, walking into the most terrific winds, not steady from one direction, I can't tell which if any direction there was. Rain was mixed with sand and that was likely mixed with saltwater. There was a lot of blurry stuff and I'm walking into the most terrific force. I didn't walk in just one direction. I couldn't walk in one direction and from this period, I kept looking, leaning my head around and trying to come to decisions, what to do, which way to turn. It wasn't like I had a pre-arranged plan and without further storm-news. Perhaps I considered the one and a half hour hike to Lihue. Just to the So. is a really pretty beach area past a "protected" curve, Cocoa Palms where Elvis married across of Wailua Beach.

LEAVES looked all black, which I just noticed as blurs flying by, within and without the clouds far above, moving in a mainly straight line. After another glance at what I'd passed, I LOOKED FOR THE THIRD TIME, by then the drive in window was such that you could Walk through it *and on through to the other side of the building and then on outside again!* The decision was made to again cross over Kuhio Highway. That was no big thing.

"WHO COULD BE CRAZY ENOUGH TO BE OUT DRIVING IN A HURRICANE" Or... FOR THAT MATTER, OUT WALKING AROUND (IN ONE) ...IT WAS INCREDIBLE TO THINK I'D JUST MET AND EVEN TALKED WITH TWO AUSTRALIANS , big and likely Surfers. These "Aussies" in Hawaii are mainly a tall lot, I met thousands, So I then thought, but the burning questions were "WHERE'S SHELTER and ... a light for a smoke?"

JUST THEN I saw a car rapidly approaching town from Lihue, A red sportscar was coming towards me, I glanced to my left, the windows of the laundromat were all gone. This was the same place that "DI" and I did all our laundry. Now it was a very torn down mess with tiny shards, splinters of green and clear glass evenly spread up the driveway and all the way back to duplex apartments facing me.

CHAPTER NINE, CONT.
LAUNDROMAT ALOHA HOSPITALITY

Jumping over a series of downed power lines, without touching any thing at all, it occurred to me I might have been electrocuted by some of the "High Juice" yet then it was too late anyhow, so what! You get a feeling of oneness with all by going over, triumph over these "things" and I just had to keep on going, feeling the terrific winds guide me on along, this was a wind of change likely not to be forgotten a long time from now

IT SEEMED like the thing to do, so I waved at the driver. She was pretty, smiling and a welcome sight to see as she made the turn into the driveway. I can't figure out to this day how she had the nerve to cross the power lines, yet I'd just done it,

When she rolled down the window, I asked her for a light and figured, she'd hand me the in-car lighter after she pushed it in. Instead of that, she waved in the direction of the apt. bldg. In a moment a big Hawaiian guy came running up to us from the second floor of the apts. He looked the essence of power and smiling confidence.

THEY SEEMED to be talking and I stood back, feeling the rain, hearing really nothing I could distinguish, it was all the effects mixing together by this time, wind noise, flying trash, cardboard boxes from across the street at the supermart, and I was pressed to concentrate on any thing. I was ignoring them and glancing to my left, became fascinated momentarily with soap boxes, and shingles, boards and clothes, flying this way that way some to Japan and all swirling around, lots of colorful debris by the tool shed and all these telephone and power poles, with the grounds littered a tremendous splattering of bits of broken glass.

The Big HI. Guy came running back down the staircase and then turned, he was smiling at me. He waved his left arm around and motioned to me, I walked over to the car, the woman still had her window rolled down in the raging winds and painful, piercing power of the lashing rain. There wasn't so much direction to the wind, it was just swirling around ... or so it seemed then. I was looking around more in a daze.

"This is for you", he gave me a red cigarette lighter which appeared to be new, one of the disposable types which coincidentally fit inside my silver case. I was pretty amazed and know I smiled, he couldn't possibly have seen me with the wind, turning my head back and forth to get the water out of my eyes. I made some kind of thankyou gesture and turned away from the winds, then I started at a fast clip back towards camp. But how he found the best thing I could have been given, I was amazed and happy at that gift. The guy had run up and then back down two sets of stairs, what speed and power!

Under these conditions of wind, it was amazing just the generosity. Very much so the hurricane was still peaking and to speak and be heard was a feat in itself. Another thing , I thought it took s a fair amount of courage to open her window in the first place, but that's Hawaiian style for you, storm or none! To this day, I recall how small, petite and pretty she was, the guy was BIG, some of his steps he swaggered from side to side and that too may have been the erratic winds.

This event of getting a "Lighter" actually brought me mentally back by association, closer with an earlier meeting of two "Aussies" due to the shock my mind was in and trying

CHAPTER NINE
LAUNDROMAT ALOHA HOSPITALITY

to relate this experience in the context of all the confusion. c This was a kind of time-warp I still can't explain after almost ten years, never claim all is logical! TO BE point blank, by the way these two Hawaiians treated me, a stranger caught outside wandering in a hurricane, I was flabbergasted! Likely I stood and smiled at them and turned back facing North and at that time, there was less wind to face. Damn if I recall thanking them, maybe I did

THEY ARE pretty direct, a little primitive and wild in their priorities yet in day to day affairs, the Hawaiian REAL people are very gentle, practical, generous and helpful.
"We Hawaiians like things a little primitive!"
Hawaiian airline employee on Maui, commenting on their new airport, with no walls !

ALSO IN KAPAA, I recall an unrelated incident but parallel in many ways, revealing their character and desire for simplicity, honesty. Again this was a real "Kane"(Hawaiians are big people) guy named Arthur who'd built himself a reputation for being honest and straightforward. "Locals" in town described his behavior like this "He's crazy and the blank-blanks don't dare mess with him." I heard that ...yet sensed Arthur WAS A PRETTY FRIENDLY GUY who just wanted to be respected instead of pestered. Besides that, he must have had some good reasons for being angry , likely some form of abuse from authorities who take people like ART as just another dumb native. I sat down and had a beer with him and it was evident he had been caught up in some kind of drug traffic, this adopted, Howle Plague is getting worse among all islanders but especially in the cities. I spent a half hour with Art and never saw him for months.

It was a hot day, probably in December, without the usual Tradewinds, which makes the heat get high and you really feel the sun of subtropic Hawaii and... I'd been chased away from my camp, then sent to court for trying to sleep on the beach, then the D. A. goofed up! You get totally sick from all this harassment, later I was hospitalized in Lihue for exhaustion then had my left arm injured for months by a nurse who was incompetent with needles.

In the soul of our big Hawaiian buddy, Art, he had the same problem, someone was really bugging him, Art just picked the guy up... and threw him away. Very direct action and although I don't approve of it, this certainly looked like something he could do.

It was a few days later Arthur had another ordeal and then suffering from mid day heat thirst, he asked if I'd give him a sip "Of Your Soda", I felt humbled, honored and gave him the whole can and would have given more had I been able or requested. People like Arthur are important to Liberty here as everywhere. I know I'm speaking for the people of all Hawaii when I write this! Over a period of time, I got to feel more for and like... the native peoples here, that is ...the For Real Hawaiian People. I have often wondered where the name "KAMEHAMEHA " usage originated, because in real life the Hawaiian People's aren't "LONELY HEARTED" (THE LITERAL TRANSLATION) . THEY TRY TO LIVE and Be Big Hearted.

CHAPTER TEN
HAWAIIAN INSIGHTS

THE HAWAIIAN WORD "INIKI" means "The Piercing Wind". I wonder who has license, makes up these various names for storms, what have you, etc. Is it N.O.A.A.? Who is given this honor and does it relate to the actual event or is this a kind of "shot in the dark" deal? ONE THING FOR SURE, this "storm" around, it was a very uniquely HI. Name perhaps invented solely for this incredible storm. It's worth mentioning HI. feats of bravery and strength during emergencies are legendary and time honored. This storm was to be no exception... but the Howle press was too busy to cover this sort of thing as was so duly, unreported for this event.

No one I ever met had heard of the HI. Language roots going back to possibly, Peru, So. America, My studies kind of "trickled down stream to the sea" after checking out Tahiti, Polynesia and Samoan influences on the HI. Language. Latin is considered a dead language. The usage of Hawaiian isn't dead but it took me about 6 years before I heard a group of women speaking it smoothly, fluently and that occasion was in the main museum on Kauai'I, in Lihue, by a couple of HI. "Waihines" (Women) who spoke in soft near musical tones, something rare, beautiful and seldom heard. You daily hear scattered words, a phrase and then bits and pieces of this disappearing tongue. THE BEST SOURCE IS CONSISTENTLY IN THEIR MUSIC ON KCCN RADIO, "STATION ALOHA" FROM HONOLULU, come to think back about it, I believe the last station I tuned to was KCCN prior to INIKI. The Sony Portable was totally swamped, buried in the muck along with my microcasette under mud and two-three ft. of sea-saltwater. I had no access or difficulty getting to a radio and water for months, NOT TO MENTION A HUNDRED OTHER THINGS including the more mundane stuff, toiletries and usual means of cleaning up! This was a living Hell to try and live with, I guess that's what the authorities had in mind, the method to their "civilized" insanity which was then and became a more highly coordinated effort to drive everyone out of any campspot. In a way , this crude treatment was for the better because (I still had about a year more of this crap to put up with and it got worse every day... with few exceptions) in the long run, I realized to my delight that I'd been relying on a lot of conditioning to standards which were unnecessary and quite artificial, that is I realized I simply didn't need a lot of crap I thought I did! With the exception of a brief visit in Hanalei Bay, where the showers were in operation, staying clean was most difficult of all, except for daily ocean swims.

Readers may ask questions like the obvious following ones;
What did the Emergency workers, proper Authorities do to help with shelter, for you?
Over a period covering more than a year, They did nothing, other than a cheap tent given, it lasted less than a month. I did contact dozens of the various bureaucrats on behalf of I and my fiancé, "DI", who was later raped at Kalapaki as a direct result of being without shelter. There on Kauai'I , both disabled!

CHAPTER TEN
SOME HI. INSIGHTS

Readers may ask questions like the obvious following ones;
What did the Emergency workers, OR ANYONE, proper Authorities do to help with shelter? Over a period covering more than a year, They did nothing, other than have me fill out forms and later a very cheap tent was given by one of the church groups, these are the real "old Money" groups that control Hawaiian Lands. That tent leaked from the first use and it lasted less than a month. I did contact dozens of the various bureaucrats on behalf of I and my fiancé, "DI", who was later raped at Kalapaki as a direct result of being without shelter. There on Kauai'I , both disabled!

I and my fiancé were "chased away" from many camps and even TEMPORARY "rest", including the beach and will mention in a later chapter about our 3AM experience.

The Federal Emergency Management Agency (F.E.M.A.) managed then by Ron Doudini, would give NO Help, actually delayed me and...WERE RUDE. The excuse or basis for denial of help was that I wasn't a resident! At the Lihue (Near Kukui Grove, as I recall) office I got a scribbled yellow carbon form, so crudely done (on Purpose) that scarcely a word of their entire form-denial for help was comprehensible!

What did my family do to help? I asked three, most of my immediate family are deceased. No help was forthcoming from any! I did get help from various friends (most notably Tod and Wanda) and acquaintances on Kauai'I, in retrospect, this was the greatest source other than the modest checks that came from disability, which I should add... were no where near enough for affording shelter of any kind.

WHAT COULD I say I'd learned of HI. People and their ways. from living here the years since the fall of 1989 ? The realization of many things , these "material goods" are all dated and can be taken and also replaced. I knew all this but still, in the realm of realization... it was living here in Hawaii and then enduring the hurricanes effect, loss of all which brought me to this realization in life. Yet something's endure, of quality and strength. TO DRIFT FARTHER AND FARTHER from our elementary connections to The Earth, Nature and our Environment, is asking for Disaster. It was the Early Hawaiians and also the Native Am. Indians who came to the same conclusions about ownership of Land, they simply didn't believe in it and never could accept it. "YOU DON'T OWN THE EARTH, THE EARTH OWNS YOU.", so said Chief Seattle. Interesting they should both come to the same conclusion because they were nearly half a world apart. This is the obvious sickness of our large cities, all over the world, not just in the USA. ESSENTIALLY THAT'S WHAT AN ELDER TOLD ME HERE ON KAUAI'I AND... OTHER NATIVE AMERICANS on the Mainland said the same of the "Great Spirits" will. Think of a hurricane in terms of an end, of human resistance to the power of natures will.

CHAPTER ELEVEN
CHANCE MEETING WITH TWO "AUSSIES"

Authors note: The exact sequence for this episode has been lost but the meeting was clearly the strangest occurrence during the very height of the storms destructive, visible effect . I recall how, it took seconds near instantly we, all three felt the humor and absurdity of our situation.

"As clearly as I can recall, the wind was then at its very peak," (I was explaining this to "DI" after Iniki) as we had supper in a tiny Mexican restaurant near Poipu. This was to be our very last meal together, she'd been near-raped down at Kalapaki and had become an even more unmanageable partner after the Med. People had harassed her out of what wits remained. She'd been bitten by a "Port-a geese" Man a War" while underwater as a guy tried to rape and force her into submission. But getting back to the "Aussies" during the storm, the real incongruity was this innocence and arrogance of our situation, not so much an attitude, like WE could stand there and decide to turn the power of a hurricane, on or off to suit our taste or just have a cigarette, by then winds were around 100-150 MPH. I can recall no other time during the entire storm when being outside, the wind had this much power, it was a point of balance I lost two or three times and literally went rolling down the highway. I had walked on up Kuhio Highway past the shopping center and was having a harder time making my way into a wind, this was then from the South or southwest because I had to lean to walk straight down the concrete. The pain of having my eyes open was necessary and it hadn't occurred to me there might be some closer place to seek shelter than heading "blindly" south.

So we were out seeing the sights, I and these two guys just walking around on the sidewalks, sightseeing which was my original idea in the first place but we got more than we could handle. Seen in perspective, until the storm was in progress, no one really knew how powerful it would be. The natives were more "in the know" and they stayed undercover, except for the two Hawaiians I met at the laundry.

I was facing the Mc Donalds or what was left of the NO. end of it, which I saw crossing the stream bridge. It was then I saw a big blue dumpster hanging off the bridges oceanside, facing East. There were two tall guys walking towards me and because this was likely the first hour of the storm, maybe even the first thirty minutes, I didn't think it too remarkable to see people, these guys and of course I had no way of knowing what we were then in, as far as the power of the whole thing. Remember as I may have mentioned, most people were sick and tired of listening to radio reports of the coming hurricane, it was worse than anti-climatic, it had become a pain in the butt and rather than remain indoors there is a likelihood that some others (at least a few) were also out "seeing the sights".

Later I spoke to dozens of fellow Kauai'I residents who'd been doing a variety of things outside during this period, photography, sightseeing, and not the least of which partying!

CHAPTER ELEVEN
ABSURD CONVERSATION

Fighting the wind and slowly going forwards, I saw how big the guys were, very tall 6' 3 or more ...the left one must have said something to me first because I did distinctly noticed an accent which even under those circumstances of wind noise, I recognized as (likely) Australian. Incredible the workings of the human mind and unthought priorities. I have always been a student of language, but this was absurd. Yet We get a lot, thousands of these Australians as surfers on the NO. shore and I particularly liked all those I'd met. I asked the guy that spoke to me if he had a light for a cigarette.

He and then the other guy started laughing at me. My reaction was one of maybe, anger and I was wondering why the guys were laughing. We were then hardpressed by the wind to stand up, it was a kind of "hackeysack" dance w/o hackey. Balance was the thing.

"We are looking for the same thing.", it was then the ridiculousness of our situation became apparent. I later wondered if we had a light, could anyone possibly light a smoke in this hurricane wind ?

They were going somewhere North and as near as I recall, I was searching for any kind of shelter and there wasn't the presence of mind to ask them where they were going? In 20-20 hindsight it would likely have been less danger and excitement to go with them but then again, I would have more likely got some(recovery) help afterwards. So What, in hindsight! The help was 90% plus for the rich who had insurance and the homeowners with their families. I must have continued on down Kuhio Highway which was getting heavily littered with tree debris, jungle canopy vines, leaves from the palm trees and roof shingles with nails still connected and now for the first time I noticed whole trees were coming down and the inevitable black masses of entire rooftops of connected shingles, palm leaves, paper and assorted flotsam and jetsam from the Pacific most likely, wind was really lifting a lot of sand, so much mainly invisible sand but it stung. The low forms of clouds swirling around were filled with this sandy debris.

Memory of this period was one preoccupied not with safety but keeping the debris out of my eyes, going to a somewhere was reduced to surviving minute by minute and maintaining balance in the wind. I had glanced at the various structures when I passed the shopping center, twenty yards back, the best thing seemed like going on SO. farther.

CHAPTER TWELVE
SHELTER FROM THE STORM

The Wyland Whale Gallery WAS THEN one of approx. two dozen shops that were open and in operation prior to the storm. "DI" and I used to go over to the healthfood store and have coffee at the tables in a circular park towards the back of the gallery. After turning and crossing Kuhio hwy. for the third time, I headed past busted oak doors and noted the power it must have too to break these, at the entrance to what was a department store. A red blinking light was some security device which was blinking via its battery and was likely the only electric light for miles. I looked in the broken windows and again admired one of the cameras I had been "window shopping" for , days earlier. Canon always had a strong camera, that was a good selling point and it occurred to me, most of those were going to rust out in a matter of hours from the saltwater.

I walked up to the front windows largely intact and looked at a baleen Whale, in a surrealistic setting beneath acrylic, phosphorescent waters, now aglow with real *saltwater* thrown in the front window from the Pacific. Those whales might go swimming *today*, I kept walking around in a semicircle and saw the sign, chained to the wall for the sandwich stand. I never bought any food there, it was too expensive, but I'd had beer there once with some six friends. It was then I noticed some candy wrappers that led the way a few yards further to a totally open storefront, still under construction, I simply couldn't believe my good fortune. Th flow pattern of the wind, kept nearly all of it off me, because of the overhang on the roof.

The roofing may have been steel or something recently constructed, it simply stayed intact. I walked the final yards looking down at the sidewalk, someone had been eating ice-cream and there was come chocolate on the sidewalk. It was then I spotted some beer bottles, I could see and even hear them rolling around on the concrete at the entrance to the next shop which was wide open for about fifteen yards. There certainly were a lot of new "Doors". This place was like a miniature Nuremberg coliseum, due to the winds of change, it was vast and never completed but what a pile of beer bottles, a dozen or maybe two dozen rolling around by the winds power, all over the floor which had some water on it and was perfectly smooth. I saw the usual debris of the const. People, the workers had a barrel which they'd neglected to put the trash in. Not that any of this mattered to anyone, now.

It seemed to me the winds had shifted a little and as I glanced out to the West, I'd decided to stay here as long as I could last. I wisely laid down on the concrete and took off my bag, the backpack. It was all wet on the outside and searching within, I found my cigarettes and opened up the package. The new red lighter worked perfectly as I lay down in the water and lit a smoke. This could have made a nifty commercial , something strange. I could just see it ... a Frank Zappa album cover, "Blowing smoke in a 200 mile per hour SHOPPING SPREE" or maybe, "CAMIL, Kauai Cowboy smoke FOR A REAL MAN IN A STORM", at any rate I was trying to imagine how much stranger it could get, could it?

What a strange situation, I was looking out at the ruins of the shopping center, not a soul in sight, the mid day was now passing into the evening and it was steadily getting darker

CHAPTER TWELVE, CONCLUDED
SHELTER FROM THE STORM

as I laid there for maybe an hour with thoughts like, what's going on and for how long. The wind was still screeching and to the left the parkinglot was clear of all that formerly was two feet or higher. Some cardboard and tree debris, shingles, nothing stood over a foot high, swept clean and empty.

I got up and must have walked around inspecting every wall, this place was going to weather the storm, it had just been put together to the latest standards for storm and Hurricane const-safety. The reinforcements on the walls were connected to the roof with steel, galvanized fittings. I liked all this; the view of the sky was the best anyone could have asked for. I had the entire shopping center to myself, unlike only about six hours earlier. I could say it was peaceful but not quiet. The wind had somehow blown in rain, I was getting drenched for the third time.

CHAPTER THIRTEEN, NEW MORNING, AND SHORE

I awoke after a--- hour nap, to begin sorting things out. Some of my back was pretty sore in the muscles and I'd been hit but never could figure out by what. It was a really beautiful (very early in the morning of Nov. 12th.) day and felt pretty good about everything. The air was so immaculately clean, you could actually taste it and it helps to know that you share your problems instead of just facing a mess all alone! This is the basic premise under which this USA started. And many civilizations have been built up to higher cultures, certainly much more durable ones: it's called Teamwork and done in a practicing spirit of Unity. I didn't expect much "Recovery help" and wasn't disappointed much. The trials I would soon face from people who were positively (backwards) a drag and simply a waste of time, were many. This morning I wasn't thinking about that at all. There was a lot of trouble to come, this I was sure of, but not yet! Bonepickers and "Allens" were soon to be arriving for curiosity and anything they could be fascinated with or carry off. THE TREES, for all directions in sight were totally stripped and the air smelled a salty, crustaceans brine, soupy something like frying seaweed or so it did by the time I got up from my plastic tarp. You got that same smell when walking the beach around One PM, when the sub-tropic Hawaiian sun was literally frying up the shoreline but this was 3 AM!

I WALKED North to the site of the old camp, it was now scene of a monumental tree uprooting, the face of the root-taproot system was up in the air and covered a large circular area. The soil uprooted with it, actually looked like some pretty good growing soil, unlike the sand just ten yards East. No way, could I get anything up sunken farther below I HAD A TWENTY, MAYBE THIRTY FOOT IN DIAMETER IRONWOOD TREE ON TOP OF IT. That was the tree I felt confident would survive and had pitched the tent behind, 2 or 3 days earlier. So much for rational preparations!

CHAPTER THIRTEEN
NEW MORNING ON A NEW SHORELINE

I had been sleeping on wet (with inland flooding, saltwater) debris, a carpet of various jungle canopy prunings, stuff like vines tangled with plastic from the (across the road) supermarket bags The sky above me was lit up brilliantly by what appeared to be ... of all things, A BLUE MOON IN A CRYSTAL CLEAR SKY. THERE WAS NO TRACE OF A BREEZE. I WAS X-AUSTED, then minutes later decided to rest some more with my storm concoction-preparation, A SIP of some very bitter "Hana Bay Rum" and O.J. to wash this whole dirty-dishwater reality to someplace else farther downstream.

The Moonlight was brilliant, (I could read a paper) the air was smelling exceptionally sweet, even by Kauai'I standards. It occurred to me that Hurricanes are natures way of cleaning a lot of Earth, affairs of Mankind, out of the way for good, more than just the deadwood from the trees. Glancing up again, the moon had turned blue, I don't know if anyone else witnessed this spectacle, but I sure did. I knew I was naturally high from lack of water, food, sleep and then too nervous bewilderment.

ON THE oceanside facing to the East beach and ten yards away I say a pile of debris which was swept in by the peak point of the waters saturation inland. There I found a heavy (very thick) in mil., plastic tarp. The edges were shredded and it made a passable blanket. The temperature was warm. As far as was visible, everything was mowed down debris.

LIKELY mosquitoes, bugs and pests had been blown clear to TOKYO OR MAYBE CHINA. There were no birds singing and a complete, total silence such as I'd never experienced was the other "blanket" for me that morning. The beautiful Sony radio was a matter of twenty yards in front of me, underwater, so no music even if I had the strength which I surely did not. COMPLETE silence reigned. Time for some Peace and Quiet, rest!. I looked up at the clear sky and felt pretty good to be damn damp and alive.

CHAPTER FOURTEEN
GETTING GUARDED GAS

"VIOLENCE IS ALWAYS THE BOTTOM LINE, BUT I GUESS THAT'S BETTER LEFT UNSAID" .   BOB .D.

    THE WISDOM of grab-it convenient power is a shaky way to rule, a kind of temporary, throw-away masking tape-guise like the thousands of yards of that stuff used to momentarily keep windows from shattering in the face of a hurricane storm, "INIKI".
    But this Mocks real strength and has nothing to do with respect, especially for peoples with the humane-ness and power of the Hawaiians, in fact, it was a total insult to everyone concerned. At the gas station near Kapaa, it was a sight no living generation islander had ever seen. People were pretty calm and laid-back, really they were for the many, and on the surface... having a good time. When you get to know the temperament here, it would be unusual and something fishy going on if the people weren't laid back and calm. Thinking about all that, I kept walking and listening to a great variety of "Boombox" music and the traditional stuff, everyone had their car radios on. At that time, there was an islandwide craze for Reggae, you could count on hearing it a dozen times an hour.
    Conversely, I couldn't count how many had rifles , wearing the para- and for real military stuff. It gave me a chill and I hope no one on the Hawaiian islands ever sees that again. I kept right on walking up to Kuhio Highway from the campsite and passed Mr. Eds pasturage, he was long gone. The thing is this, I was trying to collect my thoughts and no one appeared or was near to talk to. I had yet to meet another human since that Hawaiian guy gave me the cigarette lighter. At that point I WOULD HAVE enjoyed seeing people even if they ignored me. The group now in front had their backs to me and were facing towards a big brown -green generator set, a Gen-Set. Then, these generator goodies were in short supply. The Nat. Guard, Army and Airforce used the C-5-A , the worlds largest transport plane to fly stuff like this in from California, etc. This went on for months until these torn out utilities and infrastructure could be re-worked. A giant of a corp. began appearing, just a yellow and green trim truck at first, Henkels and McCoy Line Rebuilders of ___ had their best and likely the worlds best power-line repairmen on duty within the first week, these guys operated with Teutonic efficiency, the tab must have been in the multi-millions to consumers because ALL POLES WERE DOWN! So, later you saw a lot of widely scattered "portable" Military and civvy equipment, solitary standing stuff., but then it was the first "loose" one on the island I'd **ever seen.**

CHAPTER FIFTEEN
CLEANUP TIME

Going on up the steep hill to the left of Kuhio hwy. was the goal! I tried to get water at a hotel-motel. They were really hit bad because of their direct exposure to the beachfront. Millions of tons of sand had been blown inland and there was much small shredded bits of roofing material scattered evenly, this stuff was literally everywhere you could glance.

A woman appeared, she said the water and faucets were good! I hiked around and around until a faucet appeared. The water came out for a moment, no more. I was tired, thirsty, mad and hearing ignorant beings had just begun ... for that day. The line of people at the gas station was all the way around the ocean shore curve and backed up out of sight. So What. I continued the hike towards the NO. end of town. There seemed to be a lot of Military traffic, the "Hummers" , the minibubble Huey's helicopters overhead which looked like a weird tiny green noisy bug. I asked some other people where I could get water. Then it occurred to me, I had a terrific head ache and now my back pain was starting to kick in. Usually I kept aspirin but this time it was back in the tent under water, about two or three feet.

NEEDLESS to say as I crossed the path and drive that lead to the shopping center, it occurred that there was no need to go over there to do business, I 'd already been the very first witness to observe, that entire set of maybe two dozen shops would likely be shut down permanently or for months. It was a scene of totally shattered storefronts and splintered Oak doors as I glanced the first OAK door in front of me.

There was no clear thought left in my head, there wasn't room with the headache and back pain . I slept O.K. but was very "spaced ", I think that's the word. It was a good day to be alive!

Walking on to the hill, I was painfully dry without so much as a drop of water, I'd gone on North and finally got to the road which led up to the armory and the main bus-change stop. There were a lot of vehicles and maybe a few tents. The crowd outside was going in and out of the tent. I decided to try that. Other than my headache, sore muscles, confusion, exhaustion, thirst, loss of all I'd ever owned, loss of my wife, well... it was just another day in a re-arranged paradise. I was thinking, maybe it would have been better to just go hang around the beach and h--- with these phonies!

I grew up with the instilled respect for authority who have their "lineup" and face responsibility with the best they can do., these clowns were truly disgusting and reflected the chain of command above them!

CHAPTER FIFTEEN
CLEANUP TIME
WATER AND ASPIRIN PLEASE

Compared to thousands of other people in need, I was (POSSIBLY) too shy or my needs too modest! Yet it cannot be denied that water and a few aspirin were basic, reasonable requests after a painful, noisy, sleepless night following a major disaster. I had made the two mile hike up to the closest source of help, or potential help! The tan tent was swarmed with people and too many folding chairs. I couldn't find anyone to listen to my request for a few aspirins. A really obese lady was rude to me, I asked someone else of this "corp" of seemingly senile, elderly and obese volunteers, I was quite simply, ignored.. Likely these were on the Red Cross team which days later blew town then left Lihue and the entire island. . I never met or heard of anyone(in the fiftythree years of my life) who got the help they needed, from the AM. Red C. OR ANY other Red Cross ANYWHERE ON THE PLANET. Let a Lizzydole and Others speak for themselves!

I gave up on getting water, not to mention a shower to cleanup... after searching, walking in circles about a quarter of an hour, then walked on down a slope to another bldg., the Armory or reasonable facsimile thereof. Inside were people under strings of low-strung florescent lights. The lights were strung too damn, dangerously low. Funny a thing like that would irritate me but then the entire atmosphere was one of irritated people. The women looked like they'd just got off a lunch break, I didn't have the least desire for food and began asking the first person I spotted as a likely candidate for getting water or aspirin.

"No we can't give out aspirin", he said. This statement came as if I'd asked for something illegal or possibly dangerous. The first thought was , who is this nut? Who in hell that guy was I don't know and never saw the a-hole again. He looked like he wore some kind of wrinkled brown mask (MAYBE THAT'S WHAT IT ACTUALLY WAS) from Halloween . Things had begun to get hostile and more strange. I registered this fact within as a future frame of reference. There had to be a lot more of this to come.

Towards the back I spotted a woman who stood by white cabinets and I believe there was some medicinal apparatus. That looked good to me ...and ...she actually looked awake and Humane, unlike the thirty or so "volunteers" I'd interviewed with my eyes and been turned down and off by. They seemed very much tired, likely they really were! Yet in all fairness, it was early for everyone, I'd guess it was like about 9:30 or an hour later at latest.

I asked this woman for an aspirin. She began to ask some superficial questions and said things so totally related to the immediate surroundings, I was getting more irritated and simply... suspicious! And once again, I reminded myself of the gravity of this situation. I mean , she didn't know me and likely I was dirty (a little) and among ten thousand other storm survivors. She reached down to her left into a wooden drawer then she had a paper bag or envelope, it occurred to me that emergency and terrific pain notwithstanding, things had to be done with Military etiquette, protocol, decorum, sanitation and ... white paper bags. She took this thing whatever it was and dumped in about eight aspirins.

CHAPTER FIFTEEN
CLEANUP TIME, CONT.

I smiled, said thanks and asked about water. This was pretty damn simple! It took another quarter hour or thereabouts and SOME OTHER WOMAN MISDIRECTED ME AGAIN., all about one (My best guess is they were waiting for a delivery of water) small glass of water. Gulping down ONE aspirin, I decided to save the rest after all that hassle. Leaving and walking the two miles back to camp, I had no water. Someone finally gave me some water and that was a Hawaiian refugee of the same storm as I. The thought of going for any more ambitious thing than that was getting absurd and I was pretty disappointed trying to put these "events" in an order of sorts. If this was Emergency Assist, what would happen later on when things got a bit drawn out and people en-mass really got demanding. I'm a very, perhaps extremely gentle person, 90% and plus a pacifist. The native people here are gentle but certainly no ass-kissers.

The day was warming around ten and trucks could be seen buzzing NO. from Lihue where they were being unloaded from planes, mostly military stuff, "Deuce and a half's" crammed with guys and tree cuttings. Buzzing and that sharp staccatto of "smallbore" chainsaws were all over, islandwide. I kept on walking South past the campsite. My headache went away and in about three hours, the warmth of the sun was upon that part of the island, for how long...?

Lots of scattered debris littered all along the road in wind strewn piles, contrasted with millions of brilliant flower, Hibiscus petals of red, the bushes of which line the upper side of Kuhio Highway. A lot of immature coconuts, bright green were seen on any and every green grass area. I was now on the oceanside and came walking up to a really big apartment complex, colored a dull sun-fried green with a monster swimpool in the center of a courtyard and the patio porches facing from four or six stories above. Few people awake though, that always amazes me, they spend and save for that for years then after their arrival, they sleep! Garbage littered the pool, vine and the usual shredded jungle grunge.

CHAPTER SIXTEEN
LEARN TO FORGET

THE GATE was always open, I'd noticed it because of walking by the place on the way to the Library, this was our camps nearest Hotel, and a woman invited me in when I asked for directions and about water. I thought that was the smart priority but had a lot of other things I could have said. She wasn't very graceful, more a gatherer of souvenirs, so I saw around the room. A first-vision glimpse of the variety of debris from our hurricane, old pop bottles and floats from fishnets, she had lots of the usual Milk-case containers to put it in.

CHAPTER SIXTEEN
LEARN TO FORGET

I was learning to forget the warm rotten seaweed mixed with tree limbs that became locally scattered garbage. Here and there was a dead cat or dog that hadn't a chance outdoors. Maybe they were inside a garage or sitting on a common (prior to Iniki) Hot Hawaiian Tin Roof and watching paradise unfold until they were relocated. Death was mighty damn quick for these creatures and likely painless.

I found an ancient, very large Coke bottle and a W.W.II leather combat boot, one blown in from Nipon, Japan and the boot likely from Pearl Harbor or thereabouts on the Isle of Oahu, it was a "Western" style shoe and fairly large, military issue. The coke bottle was more than one litre, until Allen or some shocked Howle ripped it off, likely it wound up in Honolulu or maybe back in another mainland city antique shop. That thing stuck around long enough for a picture to be taken of it in front of my desk, then the P.I.S.S. invasion began, they tore my desk off the side of the tree (remember the one I built prior to the storm, it had survived the hurricane) and took everything of value, that they didn't destroy or wasn't already broken up. Than giant Coke bottle was the most beautiful green, the sun and the sea do something to "bottle Green" glass containers, it was my best souvenir, but then I had a dream "Don't Bother With the Souvenirs" and things eerie though they were, I genuinely began to understand the simple priorities of survival amidst various levels of savagery and animals of all descriptions. The worst were the "aftermath" guys like Allan, because they were so capable of violence. A different creature was "Bart" of Princeville, he was a sub-teen who made my life miserable with his destructive and thief habits, he was a rare bird after the storm but still present.

CHAPTER SEVENTEEN
PARTYTIME IN KAPAA
A HUMAN COMEDY OF MANY PLAYERS

IT WASN'T just me in a state of shock and there was a very typical answer to this Islandwide situation. I decided to hitchhike into Lihue, it was around ten in morning and then I discovered to my great delight that the "Slaveway" store in Kapaa was open for BIZ and the money machine was again up to its evil paper capacity, the Jacksons came flying out. I then discovered to my chagrin, other Hawaiians had thought (but way ahead of me!) of the same thing I had, beer! There was no more beer, anywhere to be had in the stores (of those open) in Kapaa, not a drop and anyway it would have been all warm or Hot! So I left town! There was one other Phillipino store, I wandered in dang ME, they had about 4 Millers Beers!

CHAPTER SEVENTEEN
PARTY TIME IN KAPAA

I took off on the "nail of my thumb" for Lihue, the Capitol of this "PILE OF ROCK," and then..stinking island. In minutes, some silly, sorry "Iniki Instigator of Insurance" guy dumped me off at the "Big Slave" store, there was no beer to be had there either. I turned and was pretty "Bummed Out" but still determined to look around some (where?) more. I walked about a half mile with a mouth that fantasized only beer, the fragrant smell of hops and visualizing Women again in Bikinis, dreamily brushing sand off their legs...and realized, I'd tried every store in that part of town. There was a goofy place I'd visited once, down Rice St., past the Mayors Office (in practice, this was really the Seat of Govnt. for the entire island) but I flashed on that "Montana" honkeytonk place for an instant and then started back the long hike to Kapaa, a long walk because the road past the curve leads to miles of a valleys curves and no possibility for a ride, not even on the rare occasion of a bus passing, no stops! Where's the party, I'm asking myself as I passed some office, then the split up wood, shattered ghastly remains of what had been the classiest theatre on the Emerald isle, the plaster fountain was still somewhat intact and I took a penny out of it, sat down in the sand and splinters to smoke a rolled cigarette, Damn the simple things are good!

The town was ghost dead, not that Lihue wasn't always mainly dead, it's the business capitol, figures! A Philipino guy no bigger than a sixth grader walked into a small store and the door was open, this place was actually OPEN! I had a pile of money but no place to get beer on the entire island, after asking a hundred people, all of a sudden this guy walks out with beer. I walked in, to the right was a cooler, a warm cooler. This thing was full of Miller, in sixes, I grabbed an armload and continued to the clerk. The Price was the supreme one! As I recall it, these guys were Greek or Italian and they just jacked up the price as high as they wanted. I won't mention any more about that!

So, rapidly walking out the door I felt screwed, confident and then I cracked open a Miller. A guy stopped and in a Flash of his crummy, Pacific salted-rusted out Ford, we were back at the "Mahu" palace in Kapaa. I felt like I could go for something wet and colder, much colder. The air temp was around 109 F. or maybe 119 and this place was the right everything for a heart attack or simple heatshock (if you're ever in Kapaa, it's on the Right past the second hotel but still close enough in conveniently from the Hospital) and was crawling like a wormcan with weirdoes and a hundred others sucking up the islands reality-relief, the very last of it, except for the Pocololo of course; if that be your poison.

Above the bar were the colorful trademarks of the planet for "Brewsky" as the natives here call this staple, lined up in rows were the labels of every single beer, brew, wine, whiskey and any other liquid depth you could plumb. This, however was the end..........the end of the line, for the island because after they sold all of this, there was no more of nothing, with the name beer on the label, although liquor was around and few wanted that because it's not really Hawaiian style in practice.

PLIGHT OF THE ALLENS
Or BEING UNGRATEFUL FOR LIFE, ITSELF

    I first met Jim Allen (that's not his real name) when he came to our camp from his "spot" closer, within a matter of yards of the roar of the Eastern shores of Kauai'is surf. Even the location of his tent was a clue as to his character. It's smart to be close as up to the shore-waterline as you dare get, depending on the Tradewinds... if you want to keep the mosquitoes under control, that's all automatic and well understood for anyone in a correct state of mind. His situation was that under any near full-moon he might be washed out to sea for several reasons. He was set-up in his tent, way too close to the breakwater and that would likely have been the time to fine him so intoxicated he couldn't help himself.

    There was a guy I later met in Honolulu which reminded me very much of Allen, Wayne was likely part real Royal HI. Lineage and He had a similar laziness induced more via genealogy than drink and cups; he put it this way "If you are tired enough, you can sleep anywhere", I used to see him sleeping near mid-sidewalk in Chinatown, Honolulu, during the very busiest of hours, it was incredible how the gentility of these other Hawaiians and Chinese peoples (numerous nationalities) dealt with him. Some or most got to recognize Wayne and simply (incredibly enough) became more and more careful of <u>where he would likely be</u> and walked around him! It could never, ever happen on a "mainland" sidewalk the <u>way</u> it happened to Wayne here!

    But getting back to Allen, he'd had it really close, several times only that's those <u>instances I noticed.</u> He had his tent so close, it was being washed by incoming tide for days, then I mentioned it to him again. Likely he was intoxicated or tired or both so he just smiled at me and kept walking towards his yellow tent, after he asked me for a cigarette.

    Days passed, he threatened me around One AM, after I couldn't give him some drink. Because he was a hurricane survivor, everyone on the island would have been more considerate and ignored his abusiveness, but his overall health was going down, getting worse and I sensed he had to change or else. I walked by his yellow tent daily and usually carried away some of the bottles and other trash. On one occasion I found about forty cents and gladly took that as my "Sweepers Wages" after cleaning all his debris, there was always more, the next day. If as an intrinsic and totally viable part of his otherwise unresponsive conscience he was trying to be a deadweight or it was just the hurricane shock, I don't know but this much I know, He'd threatened my life.

    In the heat of the afternoon I saw him in contortions, he was having shock-seizure of a type I'd never witnessed. I watched his body very carefully for a few seconds and debated what to do. There was just the two of us for a quarter mile on up to the shopping center.

    As a point of important information about me, I'd seen the local Med. People and they were just as out of tune, arrogant and overpriced as on the "Mainland", put differently... I wouldn't have called them unless it was certainly dire need.

    <u>There was something in the color of Allens face, this was a serious psychosomatic</u>

<u>Something</u> I knew a little of from study, **years ago** in Freiburg, GM. But not enough.

PLIGHT OF THE ALLENS
CONCLUDED

I AGAIN started speaking to him as calmly as I could and asked what he NEEDED, wanted, also I told him to try and lay still and above all quit talking. Over and over he said "I want My Mommy" which was his kind of primal scream. I was a friend and insisted, he'd be OK if he'd just Try., to relax, quit hollering, kicking, etc.

Bob his bud, came along and then called the Med. Guys, and in twenty minutes they came running down the path with a variety of technical things. The main guy in charge plugged some things on Allens triple-nippled chest and made a few readings. I believe he was a Philipino. The second medic was watching and kind of cheering the first guy along like they were all sick and maybe it could be a case of humor to the rumor. I felt all of that about an hour earlier when Al originally went into this whole absurd dance on the sand, pardoon me, it was IN the sand.

Well, things worked along their medical regimen, they shot him and felt him, checked his signs, he was too hyped up and was told to relax, I'd known and done all the above.

After throwing some trash on the sandy beach... kinda an offering to the God of the A.M.A. or maybe the Mobil Petro-plastic industry segment that makes this throwaway medical debris, they packed up their white presence and wailed off in a cloud of diesel smoke down the route to Lihue.

But the thing the "Philipino Medic" told me was this, "If you hadn't been here to calm him down, his hear was going so fast, he would have died before we arrived."

About six days later, Allen still hadn't so much as thanked me but he came into my tent in the afternoon, stole some of my few resources and then later destroyed even more of the same. This is being ungrateful for a lot! The above makes me wonder what happens to this kind of people... after they finish throwing away this life, where do these "Cretins" (as Mary Beck, the court reporter in Wanda Woods House, referred to this kind of man) deserve to go? I do my best and was really glad when he left with another Welshman who was nearly as crude, there was more and more violence going around every single day and I got more than my share of it, on top of all the other insanity and perpetual abuse from Govnt.

BIG SAVE SAFE SAVE AT HANALEI BAY

TIMING and location was perfect for filming a Tarzan movie or ...someone having a try at robbing a safe during the peak intensity of a Once in a Century hurricane, of incredible force. Hanalei Bay got hit Medium-hard by Hurricane Iniki, worst on the Northern exposed part of town facing the Bay where Sailboats and Catamarans scattered like a childs toys in the sand. These Catamarans, surfboards and other toys were shattered in bits and a thousand colored shards of plastic and fiberglass. But then and Today Hanalei is still my favorite place in the whole world! The peacefulness brought some mighty impressive people prior to the storm and I'm not going to list their names here but they mainly returned or didn't leave for the storm in the first place.
To visit, surf, or to drive up the No. shore and just have a view or even a cup of coffee will reveal this is the crown Jewel of the entire "Emerald Island" for pure, clean air and water, too it was the place chosen for filming 'South Pacific" with the unforgettable music of Richard Rogers and O. Hammerstein . I went there half a hundred times for the remainder of a work or play-day to go swimming, surfing, or a social call; beer with Jim Iron; the Best local Jazz Musician. Later it was the place to go and do a little writing and still later ... just get lost for the hottest part of the afternoon. There came a point when the entire island got too violent and that was the Post-Iniki-Shock-Syndrome, which was still ahead of us hurricane survivors. There was Bravery in many forms during and after the storm. What follows is one of the wilder gambit-adventures that failed.

Marty saw and heard a lot from his vantagepoint in the center of town, Hanalei is tiny. "The guys got there near the peak of the storm and knew this area very well. Getting in the back door must have taken some inner know-how but it's nothing that most men couldn't do and I guess what was so REALLY INCREDIBLE WAS THEIR TIMING." said Marty as he tilted his head towards the door. They tried to get out the door after using the Forklift for the heavy and most difficult part of the task. The thing about having nerve, applied nerve, was the Location. The Big S. Food store is the center of Hanalei and on top of that, next to the main Post Office! They had to wait until All the Power went out, then test for that important item.

"Then they tried to do this (get inside the store and haul out the main safe) and as far as I understood it, the only Big problem was the trees. They got in the back door and got the safe." Said Marty with a Leprechauns smile. Complications set in because the hurricane was still in progress. It was still dark out and the Hurricane made even more mess than the guys expected. Hanalei Bay is near the edge of a massive Taro Patch, lots of Palm trees were blown down and lay in the streets. "With a truck or whatever, it may have been a forklift, they got the safe out the back door and actually were going down the road with it. There were a number of unexpected things in the way, like Trees and other debris." The safe was found down the road about half a mile, not far from the park... with a tree in front of its intended path, the following morning when the "Locals" came by to clean up the brush.

There was some planning and the Robbers knew what they were doing, everything was perfect except they underestimated the Power of the Hurricane and ...they should have brought a saw to help with the trees.

Plans are fine and in hindsight it seems incredible that only two people from the entire period of the hurricane, were brought to court for trying to take something that didn't belong to them. This is two out of approx. ten thousand, according to the news and the Media. The Other group were caught either leaving, going or later on having taken some Meat and cutlery tools from a store, I believe it was in Lihue. The judge frowned on this and they got a pretty stiff sentence to set an example for the present and the future. This is very much the way of the Local Culture and everyone expected this to Happen.

In Hanalei Bay, it was a lot different "Local" scene and pretty lame politics. The thing is that in hindsight, this event was never publicized which explains , reveals WHY many things the Majority won't hear of and also a gagged and / or incompetent local media whom are either out to lunch or waiting for someone else to buy it!

HURRICANE INIKI
ONCE IN A LIFETIME EXPERIENCE

HURRICANE INIKI was a natural phenomena and as far as the majority of people on Kauai , there were A NUMBER of serious questions they asked, soul-searching to gather what they could and make sense of it, a once in a lifetime experience. Why did it occur this way, two people died of natural causes (o.k. then ... it was Fear-induced hear-attacks) and the property damage waas in the several Billions. People marked this as a turning point in their life as the Period prior and the Period After. This was an event of change which did a lot more than "sort out Deadwood" and following the storm, there was a change in the Population. Those who weren't ready to endure another one of these had been given warning and many simply packed up and left. The business cycle of the entire island went slower for years following Iniki because quite simply a number of business wouldn't move back; the Owners had simply never stopped before to realize this happens on a cyclical 20 yr. Basis.

The AM RED CROSS (NOT ONE OF MY FAVORITE GROUPS) made a recent (6-25-99) statement to the effect they are expecting more of these type major Natural Disasters, soon likely because of the changes in the Global weather system resulting from cutting too many trees, etc!. Yet even for the Hawaiian Islands, whom are visited with Hurricanes about twenty years, this was a storm of power, no living person-witness could compare it to anything, some people remembered Hurricane Ewa, however this was a much weaker storm and property damage was much less.

I DIDN'T really ask to be here FOR THIS TO HAPPEN TO ME, nor did many others yet I choose to stay... close to where I was and remain outdoors to witness it, the eyewitness "Experience" was the result and my reward . It's really a shame the photography was stolen along with so much documentation yet that's par for the course, with the general elements dealt with. THE HI. Islands are very cross-representative; that is a "Microcosm" of the whole world (The Hawaiian Islands have more than 80 languages spoken) which is most evident in Honolulu, but you hear , see it in a smaller community like Lihue and Kapaa too. This I have always felt as opportunity . After INIKI, I used this opportunity to go about visiting; interviewing and sampling public opinions on Kauai, from the North shore Hanalei Bay, all the way to the Dry Cave and then down around So. to Poipu, Koloa areas interviewing survivors with their experiences to relate. I sifted through a lot to gather (do) this work. The spot I found myself put in was simply an Outdoor. eyewitness observer, not the least a Resident; posed to react and later Witness how the "Powers that Be" dealt with this!

JUST HOW The larger society behaved under conditions of VARIOUS breakdown (water out, no electricity, some gas, few stores stocked and open, debris covered roads clogged) from this Natural disaster is the crux of the matter. Simply stated, these Islanders did very well. Without about half of the Federal Gov. and almost none of the charity groups, it would have worked out better. If they had just handed out what the people really needed instead of the paperchases, etc. I say that but the National Guard wasn't part of that group; to draw a line of distinction. These other "MIDDLEMEN-MONGERS mainly wasted a lot of money and busy peoples time, generating needless confusion.

I don't think that "On the Mainland" people would have been nearly as "Humane", well behaved, considerate and in retrospect after the fact, so COOPERATIVE IN HELPING ONE ANOTHER! This is the essential lesson, which the various layers of Government noted well. Millions were spent, maybe a lot more but the reports belched out from various agencies like the Forest Service (as one of a dozen examples) were mainly self aggrandizing and self supportive. Another example was the AM. RED Cross, their main claim is they "aid the Military" as if that isn't already overly done by the taxpayers trillions!

Each Military branch did their thing, (which is a very inefficient and competitive means to the provision of overlapping services) the Marine Corp set up Portable showers, The Army provided electricity and proved they needed new Generators and perhaps their major contribution was air-shipping of passengers out and supplies in. Every bureaucracy took notes on how to function as a better "Wall-Rebuilder".

2-22-99

SHORT STORYS OF HAWAII

- SHOOTOUT AT THE HANALEI GOURMET -

INTRODUCTION; NEAR THE END OF THE ROAD, ON THE NO. SHORE OF THE GARDEN ISLE IS THE SMALL VILLAGE OF HANALEI BAY, BEST KNOWN FOR GREAT SURFING. THEN THERE ARE THE TOURISTS WHO STOP AT THIS TINY TOWN FOR ITS CHARM, AND A BEACH ON WHICH "SOUTH PACIFIC" WAS FILMED. THE NATURAL BEAUTY OF THIS RELATIVELY UNTOUCHED SHORELINE HAS NO EQUAL FOR CLEAN WATER, AIR AND IT IS USUALLY A PEACEFUL PLACE TO SPEND A DAY. NIGHTS THE "KAUAI COWBOYS" PLAY MUSIC NEAR THE TOWNS TINY SHOPPINGCENTER.

THE TOWNS CENTER OF BUSINESS IS ABOUT HALF A MILE FROM THE COURTHOUSE AND ANOTHER HALF MILE TO THE TARO FIELDS WHERE A VAST VALLEY IS FED BY TINY, ANCIENT VOLCANIC ROCK STREAMS FROM THE LUSH JUNGLE MOUNTAINS INLAND AND BACK TOWARDS BARKING SANDS CLIFFS. THE REAL CLAIM TO FAME IS IN THE SURF AND SURFERS, SOME OF THESE BOYS ARE FOURTH, FIFTH, SEVENTH , EIGHTH GENERATION. AMONGST THE HAWAIIANS AND "LOCALS" THERE ARE EVEN MORE EXPERIENCED SURFERS GOING BACK TO THE DAYS OF HI. ROYALTYAND KING KAMEHAMEHA... THE GREAT.

AFTER SOME LONG SETS OF WAVES AND A TIME FOR LUNCH, THE BEST PLACE IN TOWN WAS ALWAYS "THE GOURMET", DIRECTLY ACROSS FROM THE EQUIPMENT OUTFITTER, JUNGLE BOB. WHO HAD A REPUTATION FOR BEING A WITTY, SHREWD BUSINESSMAN DEALING IN THE RENTAL AND SALES OF ANYTHING FROM BOARDS TO BROAD INTERIOR-JUNGLE MAPS FOR THE BACKPACK CROWD.

ON THIS PARTICULAR HOT AFTERNOON, THE COOK HAD A MINOR DIFFICULTY WITH THE CHIEF DISHWASHER, ACTUALLY IT WAS JUST A LOT OF BUSINESS FROM THE LUNCH CROWD WHO TEND TO "WOLFPACK" IN AROUND ONE AND MANY STAY TILL THE HEAT DIES DOWN OR THE TRADEWINDS PICK UP, LONG AFTER ALL THE SURFERS ARE BACK TO THE TOWNS PARKS AND OUT ON THE OCEANFRONT, FACING TO THE NORTH.

THE DISHWASHER HAD PILES OF DISHES, SOME WERE BROKEN AND GOING ROUND AND ROUND IN THE DISHWASHER, THE DRAIN HAD PLUGGED UP AND ALL THE KITCHEN AREA BECAME HOTTER AND STARTED TO SMELL LIKE SPAGETTI BEING BOILED FOR THE SECOND AND THIRD TIME, WHICH IT WAS!

NOW THE DISHWASHER WAS SAYING HE'D QUIT BUT THE BOSS FOR THE AFTERNOON TOLD HIM, " NO, WE ARE BUISY, TOO BUISY" AND IN A TINY TOWN LIKE HANALEI, THAT SORT OF NEWS SPREADS ABOUT ANYBODY WHO QUITS.

AROUND TWO P.M., THE DISHWASHER STARTED THROWING DISHES AT THE KITCHEN DOOR AND THEN SAID HE WAS GOING TO MURDER _____!

CROSSING THE STREET, WITH HIS APRON STILL ON, THE HEAVYSET DISHWASHER ENTERED "JUNGLE BOBS" AND ASKED THE TINY "WAHINE" IF SHE HAD KNIVES FOR SALE. AN ENTIRE RACK WAS SET ON THE COUNTER AND THEN HE SAID TO THE WAHINE, ' I WANT SOMETHING BIGGER"!

JUNGLE BOB HEARD THE DISHWASHER AND RECOGNIZED A TROUBLED GUY WHO WAS JUST ASKING FOR HELP?

"WE HAVE SOME REALLY NICE MACHETES , THESE ARE FROM BRAZIL AND HAVE A CASE WITH EACH ONE" SAID BOB AS HE SMILED AND PULLED ONE OUT OF ITS CANVAS BAG, PLACING IT SQUARELY ON THE GLASS COUNTERTOP. "IS THIS ONE BIG ENOUGH" SAID BOB, AS HE SMILED AND REACHED FOR THE TELEPHONE.

"YEA, THAT'S BIG ENOUGH, AND I CAN PAY YOU TONIGHT, O.K "
SO BOB SMILED, CALLED UP THE TOWNS ONLY "OFFICER" AND ALL THIS WAS HAPPENING AS HE HEARD THE DOOR SLAM BEHIND THE DISHWASHER. THE TINY WAIHINE CLERK HAD BEEN HIDING BEHIND A CURTAIN, SHE CAME BACK OUT AND TOLD BOB SHE'D LIKE TO GO HOME AND STAY THERE FOR THE DAY. "O.K." SAID BOB.

NOW THE CLERK AT THE HANALEI GOURMET HAD SEEN THE DISHWASHER RUNNING BACK WITH THE NEW MACHETE AND RECOGNIZED IT, HE HAD JUST BOUGHT

SHOOTOUT AT THE HANALEI GOURMET

ONE FROM BOB, THESE BRAZILIAN ONES WERE REALLY GOOD, VERY EFFECTIVE ON KAUAI'S JUNGLE, THAT'S THE STUFF THAT BEGAN AS YOU WENT OUT THE BACK DOOR OF THE "GOURMET" AND JUST PAST THE TARO FIELDS.

THE COOK WASN'T A FEARLESS GUY, IN FACT HE WAS OVERHEATED, STINKING TIRED, AND THEN THE DISHWASHER CAME RUNNING IN AS HE WAVED AND WEAVED THE MACHETE AROUND HIS HEAD. THE COOK LOOKED THE DISHWASHER IN THE EYE AND SAID THE FOLLOWING " DID YOU NOTICE THAT OUR OFFICER ARRIVED"?

THE TOWNS ONLY OFFICER STOOD OUTSIDE THE FRONT DOOR AS THE DISHWASHER CAME RUNNING BACK OUT THE SAME DOOR HE'D ENTERED ONLY LESS THAN A MINUTE EARLIER. HE WAS GASPING FOR BREATH AND THE HEAT WAS ABOUT 112 SUB-TROPIC HAWAIIAN DEGREES, UNDER THE SHADE OF THE ROYAL HAWAIIAN PALMS WHICH ARE ALMOST UNIQUE TO THAT PART OF THE ISLAND.

AS A RULE, YOUR "KANE-FOR-REAL DA KINE" HAWAIIAN GUY IS AND ARE BIG, REALLY BIG. THIS PARTICULAR OFFICER WAS ONE OF THE SMALLEST GUYS ON THE NO. SHORE AND HE SPOKE UP TO THE KITCHENHELPER AS THE KITCHENHELPER STARTED SHOUTING AT HIM.

"I'M GONNA CHOP UP... AND... KILL ..." SAID THE BIG HAWAIIAN AS HE GASPED FOR BREATH AND JUMPED UP IN THE AIR.

"NO YOU'RE NOT", SAID THE DEPUTY AS HE, OFFICER B------, DREW HIS SERVICE REVOLVER, AN OLD ONE THAT WAS RUSTY FROM THE LOCAL SALTY AIR. "AND ANOTHER THING, (THE NEXT TIME YOU TRY SOMETHING LIKE THIS) NEVER BRING A KNIFE TO A GUNFIGHT"!

OFFICER B_____ WATCHED THE DISHWASHER DROP HIS "KNIFE", THEN HE PUT THE "CUFFS" ON THE GUY AND TOOK HIM TO THE JAIL IN LIHUE, ON THE SOUTHERN SHORE OF KAUAI'I, WHERE IT'S ALWAYS A LOT COOLER IN THEIR COOLER..

ALOHA-

COPYRIGHT 1992,   KAUAI'I, HAWAII, USA   ROLAND L. CLARK II

BIOS:

Dr. Patrick A. O'Dougherty is the author of ten books and one of America's most highly regarded intellectual historians and activists. He is the founder of the New Abolitionists School of Diplomacy, postcolonial history, personal critical legal studies and the Green Party. He grew up in St. Cloud, MN and South Minneapolis. ODougherty attended St. John's University during the 1960s. In 1998 ODougherty was a Pastors for Peace US—Cuba Friendshipment Caravan Fellowship winner. He is an independent scholar and research professor at the University of Minnesota where he taught creative writing in General College. In 2000 ODougherty became the McNamara Alumni Center Carlson Heritage Wall 5-book Scholar at the University of Minnesota. He is also a Benedictine Oblate at St. John's University and with the Harvard Benedictines. ODougherty has recently published articles in the Minnesota Daily, the Sunflower and La Prensa. He lives in the F. Scott Fitzgerald area of St. Paul, MN.

Rev. Martin Rath. OSB. is a fifty year monk at St. John's University at Collegeville, Minnesota. He is Dr. O'Dougherty's spiritual director.

Roland L. Clark II grew in St. Cloud, MN where most of his family is buried in Calvary Cemetery. He attended St. John's University. After graduation he got involved in a military marriage which did not work out. He studied briefly in Germany and in a San Francisco film school. Clark is Slavic in background. He joined the community of expatriates in Prague where he lived for a number of years. Following this he lived in Southern Spain as an artist. After the murder of a friend of his in Tucson, he moved to Hawaii where this work originated. The work has been lost and stolen three times. It is Clark's tribute to the Aloha people. My publication of this work after a near loss is my tribute to Clark who I met at the Dorothy Day Center. Dorothy Day was one of my teachers at St. John's University. This published work came directly out of the Dorothy Day Movement and my alternative publishing company, Irish Catholic Revolution Publishing.

AFTERMATH OF INIKI: the formation of the THEOLOGY OF RECONCILIATION WITH THE EARTH SCHOOL OF ECOLOGICAL WISDOM.. The formation of the ST. CLOUD SHOOL OF ECOLOGY LITERATURE.
Hawaii is turtle island which is an original name for America. The aftermath of Iniki is the sanctificatioin of space and time of Hawaii and the Aloha people. This is the highest level of ecology literature.

This is the former home of Irish Catholic Revolution Publishing which was ransacked and forced the publication of Roland L. Clark's manuscript HURRICANE INIKI.

This is the Dorothy Day Center in downtown St. Paul where Patrick O'Dougherty met Roland L. Clark.

St. Cloud, Minnesota where Patrick A. O'Dougherty and Roland L. Clark grew up.

Rev. Martin Rath, OSB and Dr. Patrick A. O'Dougherty, St. John's University, Collegeville

Wilder Square Hi-Rise St. Paul, MN, the present home of Irish Catholic Revolution Publishing

St. Paul's Academy, St. Paul where F. Scott Fitzgerald went to school—The Theology of Reconciliation with the Earth came out of this neighborhood.

Arise Bookstore and Resource Center—Ecology Bookstore and Marketing option for HURRICAN INIKI.

The McNamara Alumni Center at the University of Minnesota where Dr. Patrick A. O'Dougherty is a featured scholar in the Carlson Heritage Wall of Books.

www.ingramcontent.com/pod-product-compliance
Lightning Source LLC
Chambersburg PA
CBHW081701220526
45466CB00009B/2843